Fast Charging and Resilient Transportation
Infrastructures in Smart Cities

Hossam A. Gabbar

Fast Charging and Resilient Transportation Infrastructures in Smart Cities

Springer

Hossam A. Gabbar
Energy Systems and Nuclear Science
University of Ontario Institute of Technology
Oshawa, ON, Canada

ISBN 978-3-031-09502-3   ISBN 978-3-031-09500-9   (eBook)
https://doi.org/10.1007/978-3-031-09500-9

This Springer imprint is published by the registered company Springer Nature Switzerland AG
The registered company address is: Gewerbestrasse 11, 6330 Cham, Switzerland

Contents

Chapter 1
Introduction

Hossam A. Gabbar

1.1 Mobility

Communities seek improved mobility to facilitate the movement of humans and other related objects that support human well-being and promote community development. Mobility of humans, animals, food, or goods requires effective planning and management of transportation technologies and infrastructures. In addition, it is based on city and urban planning, including roads and associated facilities, including parking areas, malls, and remote areas. Several factors impact mobility and are controlled by performance measures, such as population, city plan, community development, and local resources [1]. The use of transportation and information technologies will improve mobility by reducing service time and improving management [2]. Traffic is an essential factor for improved mobility, where effective traffic management will lead to enhanced mobility [3]. Also, business models could improve mobility, such as car sharing, where the number of vehicles on the road could be reduced and positively enhance mobility [4–5]. Many countries realized the importance of improving mobility and transportation and advanced research and development entities to study significant challenges and gaps and propose techniques and technologies [6]. On the other side, city planning is another crucial area to improve mobility where small and large events are directly related to mobility. Several studies presented effective modeling and planning for mega-events to ensure smooth mobility with different transportation management schemes [7]. Sustainability is another factor that should be studied when planning mobility, where greenhouse gas (GHG) emissions from transportation contribute to climate

This chapter is contributed by Hossam A. Gabbar.

H. A. Gabbar (✉)
Energy Systems and Nuclear Science, University of Ontario Institute of Technology,
Oshawa, ON, Canada
e-mail: Hossam.gabbar@uoit.ca

© Springer Nature Switzerland AG 2022
H. A. Gabbar, *Fast Charging and Resilient Transportation Infrastructures in Smart Cities*, https://doi.org/10.1007/978-3-031-09500-9_1

change [8]. Policymakers are studying sustainability in transportation with proper analysis of mobility models [9]. Transportation is also linked with the community economy, studied and reflected in policies to facilitate mobility [10–11]. Social studies also study mobility to support community and gender [12]. With the recent development of connected and autonomous transportation, the world seeks improved mobility. More studies are undergoing for proper preparation and testing of connected and autonomous vehicles [13]. Human factors are also linked to mobility planning and management, impacting human-vehicle interactions and community use of transportation [14].

1.2   Transitioning of Transportation Technologies

There is a recent and continuous development of transportation technologies. Electric or hybrid electric vehicles, including plug-in hybrid electric vehicles, are significant shifts to enable transportation electrification to reduce fossil fuel dependence and GHG emissions. Transportation electrification required the availability of charging infrastructures with different technologies to reduce charging time while improving energy efficiency and consumption. The recent development in hydrogen energy enabled the penetration of fuel cells in the transportation sector based on hydrogen production and supply chains. The development of hydrogen vehicles is also tested and compared with fuel cell vehicle technology in terms of performance and lifecycle cost. There are similar progress and development of other transportation systems, including bike, motorcycle, scooter, shuttle, rail, drones, and UAV (unmanned aerial vehicle), in addition to marine transportation and aviation. More intelligent features are added to electrification and energy technologies and management. The recent development of connected and autonomous transportation is another significant milestone of transportation technology. Research is exploring possible improvements to the transportation infrastructures, including intelligent features, smart traffic, 5G, additional safety functions, monitoring and real-time data analytics, data centers, and other related features that require further research and development. Several developments are related to unmanned aerial vehicles (UAVs) and their applications in aerospace and aviation and more developments on hypersonic air travel. Concorde Supersonic airlines were well developed to fly faster than sound (770 mph) and new development to reach hypersonic speed (3800 mph) or higher.

1.3   Transportation Electrification and Charging Technologies

The electric power networks are undergoing fundamental transformation characterized by increased utilization of distributed resources like energy storage systems and others, the widespread deployment of advanced based storage systems and their associated controllers, and intelligent monitoring and control technologies. This change has spurred the societal desire for a sustainable, environment-friendly, affordable electricity supply. In the process, many renewable-based distributed generation resources are being installed. Energy storages are the critical ingredient for integrating with wind or solar power generation units and in large-scale power transmissions and distribution systems [15, 16]. Multilevel control strategies of energy systems is emerging in the coming years. Together with current measurement units, modern technologies allow the implementation of comprehensive wide-area control approaches to optimize system operation and enhance security. Many of the ongoing research activities at the moment are, therefore, guided by the vision of an intelligent operation of power substations comprising still a large number of traditional equipment operating alongside modern systems employing local and global intelligent controls [17–19].

Over the past two decades in Canada, a long-term plan for disseminating clean energy sources to be integrated with the power system grids is a commendable solution for harvesting and managing integrated energy solutions. The recent research trend of power networks is to develop new methodologies to provide essential contributions making use of optimal power flow with voltage and frequency regulation and power system protection with complete short circuit analysis and risk assessment in an integrated power system analysis platform.

Most modern power system networks plan to implement smart infrastructures to provide high-performance energy supply to cover all geographical regions with reliable and cost-effective energy grids. In order to ensure a successful migration to smart grid technologies, it is essential to provide robust grid monitoring, controlling, and protection infrastructures and solutions. Accurate and timely feedback on the exact condition of the grid topology can be used to plan and evaluate the different design and operational scenarios of the future grid architecture. In addition, the integrated monitoring, controlling, and protection infrastructure and technologies can be used for normal operation and emergencies and effective control and protection of grid topology [20–22]. The design and implementation of integrated monitoring, controlling, and protection systems and technologies can be presented; and intelligent monitoring and self-healing algorithms for protection and control of the different segments of grid infrastructures are analyzed and discussed. Advanced technologies will be selected, designed, and implemented to collect real-time data. Many energy efficiency techniques with novelty will be provided. Real-time awareness system for self-healing and blackout will be studied.

Governments have a growing concern about climate change and global warming and aggressively consider reducing GHG emissions. The transportation sector is

Table 1.1 Electric charging

Vehicle	Charge as per 100 km	Cost per unit	Cost
Electric	16 kWh	× $0.08/kWh	= $1.28
Gasoline	8.4 L	× $1.35/L	= $11.40
As another comparison, fuel = $0.13 per km, EV = $0.02 per km			
Electricity is cheaper nine times more than gasoline electricity saves around $2000–$3000 a year per car, based on 30,000 km/year			

Table 1.2 Charging time of different charging categories

Considering 100 km, charging time	Voltage	Max. current	Power
6–8 h	220 Vac	16–18 A	3.2 kW
3–5 h	220 Vac	32–36 A	7.4 kW
2–3 h	400 Vac	16 A	10 kW
1–2 h	400 Vac	32 A	22 kW
20–30 min	400 Vac	63 A	43 kW
20–30 min	450–505 Vac	120–135 A	50 kW
10 min	350–550 Vac	320–380 A	120 kW

one of the most significant contributors to GHG emissions. Canada is seeking to implement clean transportation via the development of transportation electrification solutions. The development of new transportation infrastructures will support the electrification of transit, buses, railways, fleets, and personal and commercial vehicles.

Most of the charging platforms are on-road locations associated with electrical service networks. Others are situated within retail, commercial malls or operated through numerous privately owned businesses. Estimation of cost of transportation electrification charging is shown in Table 1.1. Analysis of charging time is shown in Table 1.2.

DC fast chargers replace Level 1 and Level 2 charging stations and are intended to charge electric vehicles rapidly with electric power between 50 kW – 120 kW, as shown in Table 1.3. Most electric vehicles can be completely charged with DC fast charge capacity, and there are an increasing number of chargers in the United Sates that are fit for charging huge range to an EV in a short time.

Fast charging has significant benefits to charge demand and distribution networks in big cities that are very congested and where peak demand reduction is a topic of interest. These benefits can reflect grid services to provide peak power by turning peak load into baseload. In addition, grid stabilization can be achieved by supporting grid expansion and maintaining promised supply levels.

Table 1.3 AC/DC charging categories

Fast charging can be achieved using Level 2 (basic) or Level 3 (ultimate)
AC Level 1: 117V 16A max (normal charging)
AC Level 2: 240V 32A or 70A (basic fast charging)
Charger specifications Level 3 (ultimate-direct DC):
Input three-phase 200V
Output:
Max DC 45–50 kW (up to 200 kW)
Max DC voltage 700V
Max DC current 750A

1.4 Challenges of Fast-Charging Station Development

There are several challenges related to the development of fast-charging stations. Given load profiles and grid conditions, technology selection and sizing of energy storage, such as battery and ultracapacitor, should be accurately specified. Energy management of the fast-charging station and buses is another area where there are several recent research and innovations to define control strategies and management schemes with optimization based on accurate modeling and simulation and evaluation of performance measures. The effects of fast-charging stations on the grid should be evaluated and reflected in the design of the target fast-charging station. To complement the grid, hybrid energy systems are required to be integrated with the fast-charging station. The design configuration and sizing of the target hybrid energy system should be carefully assessed to meet the fast-charging station's demand profile, with grid condition analysis. Integrated modeling and simulation are required to encompass fast-charging station with the hybrid energy system while considering all possible design and operation scenarios of the integrated system.

The adoption of electric buses will positively impact the grid with economic benefits. However, there are barriers to deploying eBuses, listed in three dimensions, technological, financial, and institutional, as shown in Fig. 1.1.

1.5 Summary

This chapter provided an introductory discussion about fast-charging infrastructures and their deployment within smart cities. Mobility aspects are analyzed, and needs are elaborated with mapping to transportation electrification strategies and planning. Examples of AC/DC charging stations and charging times are illustrated to explain the design and operation of fast-charging stations.

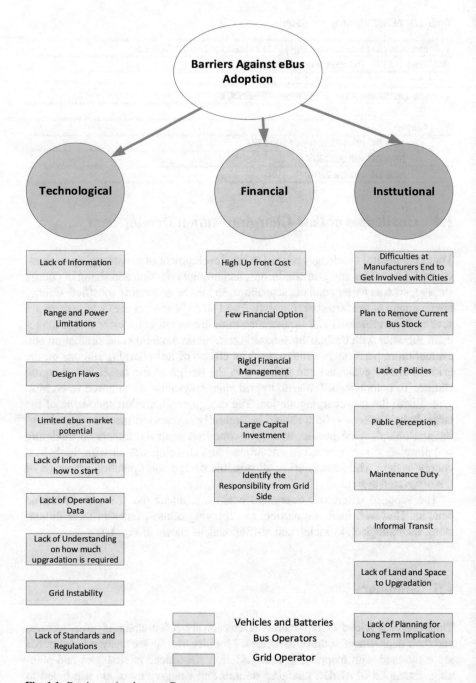

Fig. 1.1 Barriers to implement eBuses

References

1. Y. Tyrinopoulos, C. Antoniou, Factors affecting modal choice in urban mobility. Eur. Transp. Res. Rev. **5**(1), 27–39 (2013). https://doi.org/10.1007/s12544-012-0088-3
2. B. Djordjevic, E. Krmac, Key Performance Indicators for Measuring the Impacts of ITS on Transport, Isep, 2016, no. March, p. 2, 2016
3. I. Kaparias, M. G. Bell, M. Tomassini, Key Performance Indicators for traffic management and Intelligent Transport Systems, Isis, no. 3, p. 75, 2011 [Online]. Available: http://www. transport-research.info/Upload/Documents/201204/20120402_174223_62254_D. 3.5 – Key Performance Indicators for traffic management and ITS.pdf
4. T. Schiller, J. Scheidl, T. Pottebaum, Car Sharing in Europe – Business Models, National Variations and Upcoming Disruptions, Monit. Deloitte, no. 6, pp. 1–8, 2017 [Online]. Available: https://www2.deloitte.com/content/dam/Deloitte/de/Documents/consumer-industrial-products/CIP-Automotive-Car-Sharing-in-Europe.pdf
5. W. Loose, The State of European Car-Sharing: Final Report, State Eur. Car-Sharing, no. June, p. 129, 2010 [Online]. Available: http://www.carsharing.de/images/stories/pdf_dateien/wp2_report__englisch_final_2.pdf
6. E.A. Rodriguez, FTA Annual Report on Public Transportation Innovation Research Projects for FY 2019, Fed. Transit Adm., no. 0129, 2020 [Online]. Available: https://www.transit. dot.gov/sites/fta.dot.gov/files/docs/research-innovation/115886/fta-annual-report-public-transportation-innovation-research-projects-fy-2017-fta-report-no-0120.pdf%0Ahttps://rosap. ntl.bts.gov/view/dot/39036%0Ahttps://trid.trb.org/view/
7. Y. Chen, M. Spaans, Mega-event strategy as a tool of urban transformation: Sydney's experience, 2009
8. B. Hildebrandt, A. Hanelt, E. Piccinini, L.M. Kolbe, T. Niero-Bisch, The Value of IS in Business Model Innovation for Sustainable Mobility Services – The Case of Carsharing. Int. Conf. Wirtschaftsinformatik, 1008–1022 (2015)
9. C. Chirieleison, L. Scrucca, Event sustainability and transportation policy: A model-based cluster analysis for a cross-comparison of hallmark events. Tour. Manag. Perspect. **24**, 72–85 (2017)
10. A. De Oliveira, Megaevents, urban management, and macroeconomic policy: 2007 Pan American Games in Rio de Janeiro. J. Urban Plan. Dev. **137**(2), 184–192 (2011)
11. E. Kassens, Transportation Planning for Mega Events: A Model of Urban Change. Dr. Thesis, 223 (2009)
12. M. Queirós, N. da Costa, Knowledge on gender dimensions of transportation in Portugal. Dialogue and UniversalismE **3**, 47–69 (2012)
13. A.V. Notes, TAG : SHANGHAI CHINA ADVANCES PASSENGER – CARRYING ROAD TESTS OF SELF – DRIVING VEHICLES, no. June 2019, pp. 2019–2021, 2021.
14. N. Merat, D. De Waard, Human factors implications of vehicle automation: Current understanding and future directions. Transp. Res. Part F Traffic Psychol. Behav. **27, no. PB**, 193–195 (2014). https://doi.org/10.1016/j.trf.2014.11.002
15. C. N. · P. Apr 12, 2018 11:04 AM PT | Last Updated: April 12, and 2018, TransLink unveils new fast-charging battery-powered bus | CBC News, CBC, 12-Apr-2018 [Online]. Available: https://www.cbc.ca/news/canada/british-columbia/translink-rolls-out-battery-powered-bus-1.4616536. Accessed 23 Apr 2019.
16. S. Naumann, H. Vogelpohl, Models and Methods for the Evaluation and the Optimal Application of Battery Charging and Switching Technologies for Electric Busses, CACTUS
17. M. of T. Government of Ontario, Charging electric vehicles [Online]. Available: http:// www.mto.gov.on.ca/english/vehicles/electric/charging-electric-vehicle.shtml. Accessed 27 Feb 2019.
18. eBus charging infrastructure, siemens.com Global Website [Online]. Available: https://new. siemens.com/global/en/products/mobility/road-solutions/electromobility/ebus-charging.html. Accessed 23 Apr 2019.

19. New fast-charge system makes e-buses a more appealing solution than ever [Online]. Available: https://phys.org/news/2018-11-fast-charge-e-buses-appealing-solution.html.   Accessed   23 Apr 2019.
20. How did Proterra get an electric bus to cover 1,100 miles on a charge? A bigger battery, of course, Green Car Reports [Online]. Available: https://www.greencarreports.com/news/1112774_how-did-proterra-get-an-electric-bus-to-cover-1100-miles-on-a-charge-a-bigger-battery-of-course. Accessed 23 Apr 2019.
21. Charging time summary | Knowledge center, The Mobility House. [Online]. Available: https://www.mobilityhouse.com/int_en/knowledge-center/charging-time-summary.   Accessed   23 Apr 2019
22. Charging & Energy Management, The Mobility House [Online]. Available: https://www.mobilityhouse.com/int_en/charging-and-energy-management. Accessed 23 Apr 2019

Chapter 2
Requirement Analysis of Fast-Charging Stations

Hossam A. Gabbar

Acronyms

EV	electric vehicle
EB	electric bus
ET	electric truck
FCS	fast-charging station
MEG	micro energy grid
FCVs	fast-charging stations
GHG	greenhouse gas

2.1 Introduction

Transportation electrification is critical to achieving reduced GHG emissions, which support worldwide movements to deal with climate change. Transportation electrification has several directions. One main strategy is based on battery-based electric vehicles (EVs). Several challenges limit the expansion of EV. Battery capacity is an important factor in expanding the range of EV. Battery-recharging capabilities will ensure efficient repeated charging of EV and electric buses (EBs), which will support city transit service [1]. The development of battery technology led to enhanced charging capabilities and paved the way for fast charging. The mobility requirements and battery specifications of EVs, EBs, and ETs (electric trucks) will be used to establish the proper design and sizing of hybrid DC fast-charging stations [2].

This chapter is contributed by Hossam A. Gabbar.

H. A. Gabbar (✉)
Energy Systems and Nuclear Science, University of Ontario Institute of Technology, Oshawa, ON, Canada
e-mail: Hossam.gabbar@uoit.ca

© Springer Nature Switzerland AG 2022
H. A. Gabbar, *Fast Charging and Resilient Transportation Infrastructures in Smart Cities*, https://doi.org/10.1007/978-3-031-09500-9_2

The continuous development of charging techniques led to wireless charging capabilities based on power transfer approaches, such as conductive automated connection devices [3]. The requirement analyses of both battery technologies and charging infrastructures are used to design fast-charging stations. The location of charging infrastructure is important and considered as part of the requirements to establish fast-charging infrastructures and apply them on bus networks. Research showed intelligent techniques to identify optimization models for cost-effective charging infrastructure placement and optimize battery sizing of fast charging for electric buses [4]. The understanding of electric bus fleet scheduling will be reflected in the requirement analysis of fast-charging infrastructure [5]. The analysis of fast-charging station deployment for battery-electric buses should consider electricity demand charges, which will impact the analysis and optimization of fast-charging stations [6]. Case studies were investigated to optimize electric buses' recharging and reflect scheduling. Examples are shown for urban electric buses [7]. The mobility requirements are mapped to battery capacity and recharging needs for electric buses, for example, city transit service [1]. Analyzing and optimizing different components within a fast-charging station is important and includes sizing based on design and control requirements [2]. Charging demand requirements are linked to parking areas, which could be analyzed based on economic analysis and penalty models of delayed charging [8]. The analysis of market requirements and EV penetration is reflected in charging infrastructures [9]. Charging planning is also contained with power supply limits from the grid, particularly fast-charging stations [10]. Charging demand is studied for home-based EV charging, where intelligent algorithms such as game theory demonstrate an optimization model [11]. The charging is part of power system operation with direct interfaces to the grid and important charging parameters to create an uncoordinated charging load profile [12]. The analysis of fast-charging stations is highly dependent on the technology used. Fast-charging technologies, such as CHAdeMO, can deliver up to 62.5 kW by 500 V and 125 A direct current for battery electric vehicles [13]. Other technologies are specified based on requirement analysis of electric bus charging, such as OppCharge [14]. Standards are employed to define guidelines and best practices. SAE (Society of Automotive Engineers) is an important organization in the United States working on standard development. Some of their standards are dedicated to supporting charging for transportation electrification [15, 16]. Cost models are important factors as part of the requirement analysis of charging stations, which includes the lifecycle cost of electric buses with different charging methods [17]. The cost analysis will require achieving cost-competitiveness among different views within the charging infrastructure for electric bus operations [18]. The circuit design analysis for different power ratings is discussed to support charger design [19]. Power circuit design includes wireless or contactless EV charging where a number of techniques are studied, such as flux couplers with different designs and configurations, which are reflected to design requirements of wireless charging infrastructures [20, 21]. Static and dynamic inductive wireless charging techniques are studied for EVs [22]. Charging can also be used for in-motion EV, which will improve mobility [23]. Flywheel as energy storage is used for fast charging. Studies design wireless charging using flywheel energy storage systems [24]. The research and analysis of charging infrastructures revealed a number of features and innovations for future fast-charging technologies [25–27]. Charging mechanisms showed different designs of onboard integrated chargers [28, 29]. EV charging topologies with control schemes can be integrated within smart cities [30]. The charging station can offer two-way power flow with vehicle-to-grid (V2G) capabilities, supporting the balancing grid and achieving profitable charging infrastructures [31]. Charging could be implemented with an on-road charging mechanism [32]. The penetration of EV in the grid might cause instabilities. However, proper EV management and coordination can enable a smart grid with charging infrastructure [33]. EV technologies are applied on off-road transportation applications, where demand analysis is performed to support charging infrastructure deployment [34]. The technology development of EV should reach readiness levels to enable successful deployment projects [35]. The review of battery technology evaluated performance measures for EV and transportation electrification [36, 37].

2.2   Requirement Analysis of Fast-Charging Station

The design of a fast-charging station (FCS) should ensure high performance for all functions offered by fast charging. In order to properly design FCS, it is essential to perform a detailed requirement analysis to be able to provide all capabilities and functions. Figure 2.1 shows the main functions of FCS that cover the design and operation sides. Functions include managing the charging process, energy management of FCS, managing FCS facility, and the design functions of FCS. A detailed

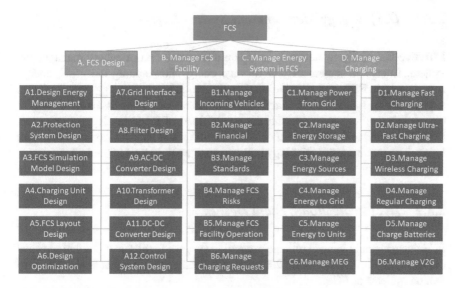

Fig. 2.1 Functional modeling for FCS design and operation

Table 2.1 Categories of requirement analysis of FCS

R1. Design requirements
R2. Control requirements
R3. Operation requirements
R4. Technical and technology requirements
R5. Physical system requirements
R6. Performance requirements
R7. Economic requirements
R8. Environmental requirements
R9. Mobility requirements
R10. Safety requirements
R11. Social requirements
R12. Standards and policy requirements

breakdown of each primary function is described. The requirement analysis of each function will be described in the following sections.

Requirements for each function are described based on categories, as shown in Table 2.1.

Each category is defined with a set of state variables and performance measures. The detailed requirement analysis for each function is defined in the following sections.

2.3   FCS Design Requirements

2.3.1   [A1] Energy Management System Design

Effective fast-charging station is based on a robust design of energy management system. Energy management will manage energy flow among the grids, energy sources within FCS, and EVs bidirectionally. Monitoring of energy flow is vital to support energy management. Management of energy storage is essential to balance energy supply from the grid with the energy load of incoming EV for charging. Energy demand is defined based on incoming EVs and their SoC (state-of-charge) levels. The requirements of energy management are described in Table 2.2.

2.3.2   [A2] Protection System Design

FCS design should include reliable protection systems for all functions and physical components within FCS. Table 2.3 shows the detailed requirement analysis of protection systems of FCS.

Table 2.2 Requirement analysis of energy management of fast charging

Component	Function	Requirements
Grid interface	Energy supply from grid	Peak shift from on-/mid-/off-peak
		Reduce frequency difference between grid and FCS
		Balance grid capacity with FCS loads (MW)
		Balance grid with local generation in FCS
		Balance grid with local generation in FCS
		Manage electricity price between grid and EV charging
	Energy back to grid	Reduce frequency difference between grid and FCS
		Peak shift from off-/mid-/on-peak
		Balance voltage regulation
		Balance frequency regulation
		Balance on-/mid-/off-peak utilization
		Balance grid with local generation in FCS
		Manage electricity price between grid and EV charging
Transformer		Support FCS target power rating
		Reduced hysteresis losses
		Reduced eddy current losses
		Real-time monitoring of transformer component health condition
Filter		Support FCS target power rating
AC-DC converter		Achieve high-power factor
		Achieve high displacement factor
		Achieve minimum harmonic distortion
DC-DC Converter		Achieve high-power factor
		Achieve high displacement factor
		Achieve minimum harmonic distortion
MEG		Achieve maximum generation capacity
		Achieve balanced energy storage to meet FCS capacity requirements in view of local generation and grid condition
		Achieve balanced capacity with grid supply to meet FCS demand
		Achieve maximum power quality levels, power factor, displacement factor, minimum harmonic distortion, and minimum losses
		Real-time monitoring of MEG component health condition
		Achieve minimum energy not served
		Achieve minimum energy not served
EV battery		Maximize battery lifetime with optimum SoC
		Real-time monitoring of EV health condition and charging condition
		Support fast-charging, with reduced charging time

Table 2.3 Detailed requirement analysis of FCS protection

Component	Requirements
EV battery	Achieve current protection
	Achieve voltage protection
	Achieve energy protection, directional active/reactive overpower
	Achieve thermal protection
	Achieve real-time monitoring
	Achieve real-time fault detection and diagnosis
FCS battery	Achieve current protection
	Achieve voltage protection
	Achieve thermal protection
	Achieve real-time monitoring
	Achieve real-time fault detection and diagnosis
DC bus	Achieve current protection
	Achieve voltage protection
	Achieve real-time monitoring
	Achieve real-time fault detection and diagnosis
DC-DC converter	Achieve current protection
	Achieve voltage protection
	Achieve real-time monitoring
	Achieve real-time fault detection and diagnosis
AC-DC converter	Achieve current protection
	Achieve voltage protection
	Achieve real-time monitoring
	Achieve frequency protection
	Achieve real-time fault detection and diagnosis
FCS controller	Achieve security protection
	Achieve real-time fault detection and diagnosis
	Achieve real-time monitoring
	Achieve protection from unnecessary tripping
Flywheel	Achieve current protection
	Achieve voltage protection
	Achieve rotation protection
	Achieve frequency protection
	Achieve real-time monitoring
	Achieve real-time fault detection and diagnosis
Devices	Achieve breaker protection
	Achieve relay protection
	Achieve thermostat protection
	Achieve sensor protection
	Achieve fuse protection

2.3.3 [A3] Design FCS Simulation Models

This section discussed the requirements of the simulation models to support the design and control of FCS. Simulation models include steady-state models, transient models, real-time models, fault models, control models, protection models, and optimization models. These models cover all subsystems and components within FCS, including grid, MEG (micro energy grid), FCS, grid filter, grid transformer, rectifier, DC-DC converter, AC-DC converter, energy storage (battery, flywheel, ultracapacitor), PV, wind, controllers, charger unit, and EV battery. Key model parameters include power, current, voltage, efficiency, power factor, phase, frequency, losses, and performance measures.

2.3.4 [A4] Charging Unit Design

The charging unit should be designed based on requirements in view of user demand and mobility requirements. Key target features include charging level, connector type, DC or AC, wireless or wired, voltage, power, protection functions, phase 1/ phase 3, charging time, and uni-/bidirectional.

2.3.5 [A5] FCS Layout Design

FCS layout is defined based on several requirements, including location, area, number of charging units, mobility demands, incoming EVs, EBs, and ETs, number of swapping batteries, PV area, wind area, storage area, control room area, store area, and entrance and exit areas. An example of station layout is shown in Fig. 2.2.

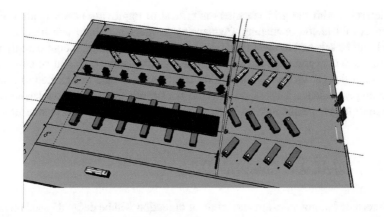

Fig. 2.2 Charging station layout

2.3.6   [A6] Design Optimization

The design optimization is essential to achieve high-performance FCS in view of input parameters and performance measures, which will be translated into objective functions and constraints. Optimization will be based on multi-objective optimization for each unit and globally within FCS. Optimization is evaluated based on performance measures such as energy efficiency, cost, time, grid impacts, and protection measures. Optimization should consider incoming EVs, EBs, and ETs, mobility requirements, grid condition, local generation and storage within FCS, and utility benefits.

2.3.7   [A7] Design Grid Interface

The main energy supply of the charging station is typically based on the power supply from the grid. The configuration of the grid interface will enable a two-way power flow between the charging station and the grid. The grid interface design is established while considering power electronics, including bidirectional converter, digital or analog PLL, virtual synchronous machine, and power management. The selection of components and their parameters is considered to meet target specifications and performance measures. Smart metering capabilities are used to ensure real-time and accurate monitoring and control. The interface via bidirectional power flow will ensure flexible operation with dynamic decoupling, bidirectional fault management capabilities, and smart communication functions to ensure real-time operation.

2.3.8   [A8] Filter Design

The interface with the grid via filters is critical to improving power quality. Filter design considers low switching frequencies to meet harmonic constraints. Filter model will be selected to meet target performance. Filter design should comply with grid-side and converter-side harmonic limits. Passive damping should be considered in the filter design. Proper sizing of filter components should be selected to achieve optimum performance. It is essential to consider systematic design, modular, and ability to meet target specifications of current, voltage, power, and performance targets.

2.3.9   [A9] AC-DC Converter Design

The interface topology design and mode of operation will be defined based on interface requirements. For example, pulse width modulation (PWM), transient-voltage-surge (TVS) diode, split-phase single-phase, three-phase, asymmetrical single-phase, two-stage, multi-stage converter, or bridge topology. The selection of power electronics and semiconductor technology is based on design specifications, target functions, and performance. Design should satisfy capabilities such as power and voltage rating, fault blocking, switching performance such as time and frequency, efficiency, losses such as conduction and switching, leakage, self-pre-charge, temperature operation range, lifetime, and physical dimension such as weight. The operation design of the target AC-DC converter is defined as part of design specifications. Fault analysis, diagnosis, recovery, and resiliency operation are defined within the design specification. Control design should cover resonant control functions, stability, transient, steady state, and error reduction.

2.3.10 [A10] Transformer Design

Smart transformers are an essential part of grid integration, in particular with charging stations. Design specifications of the transformer should satisfy target performance. Requirements include input and output power rating, voltage rating, frequency, temperature profile, power losses, efficiency, physical dimensions such as weight, protection, and control capabilities. Control functions include steady-state, transient analysis, adaptation to loading changes, and power flow control. Control design and architecture should meet the target performance of the transformer. Accurate monitoring and measurement are essential functions of the target transformer.

2.3.11 [A11] DC-DC Converter Design

DC-DC converter is an integral part of fast-charging stations. Several standards such as IEEE1547, IEC1000-3-2, and the US National Electric Code (NEC) 690 discussed DC interfaces with grid and microgrid, considering harmonics and other power parameters. It is essential for EV chargers with bidirectional power flow between EV and charging station and the grid. DC-DC converter design configurations include multi-input buck/buck-boost converter, multi-input DC-DC boost converter, hybrid buck/buck DC-DC converter, bridge-type multiple-input DC-DC converter, multi-input converter based on the switched capacitor, multiple-input DC-DC-positive buck-boost converter, and revised configuration with state-of-the-art technologies. In order to improve the efficiency under different operating conditions, dynamic inductor control is used in the DC-DC converters [38, 39]. The requirement analysis will ensure target performance measures are met, including reduced switching losses, low cost, high power density, small size, and small weight. The performance should be achieved for all operation modes.

2.3.12 *[A12] Control System Design*

Control system is designed based on requirement specifications to define inputs from sensors, output control signals, and control logic. The timing of monitoring points will be defined based on the dynamic behavior of each component and control functions. Each component's protection control requirements are associated with input and output parameters and performance measures. Control functions are defined for grid interface, transformer, filter, AC-DC converter, DC-DC converter, charging unit, and EV battery control. The control analysis covers power flow, transient analysis, steady-state analysis, steady-state error analysis, and manage load fluctuation. Control systems should consider the reliability of physical systems, such as battery wear. The placement of measurement devices within the station can be analyzed and optimized based on control loops and protection scenarios. Bidirectional digital communication is part of the control layer to ensure real-time and accurate data communications between sensors, controllers, and energy management. Integrated data center to monitor and control operation of the FCS will include real-time data, co-simulation, and history data. Data analytic capabilities will be utilized to control FCS functions.

2.4 FCS Facility

2.4.1 *[B1] Manage Incoming Vehicles*

The management of incoming vehicles (EVs, EBs, and ETs) to the station is based on a number of charging units. The constraints of maximum waiting time, charging loads, and station charging capacity are used to manage the scheduling and operation of incoming vehicles. The analysis of charging load profile is based on time of the day, any day of the week, and day of the month, while considering controllable and uncontrollable factors that affect the incoming vehicle. Traffic around the location of the station will impact incoming vehicles.

2.4.2 *[B2] Manage Financial Model*

The financial model of the station is based on the income and expenditure model. Incoming costs include charging income and the price of power returned to grid. Expenses include facility and lease costs, capital, operation, maintenance, energy supply, and less cost. Lifecycle cost analysis is essential to accurately describe the financial model of the station. The financial model will require accurate mobility analysis to estimate expected revenue while predicting generation capacity and price. Peak shaving and shifting on-/mid-/off-peak will enable FCS to profit. Similarly, FCS can profit from power returned to the grid by shifting off/mid/on the peak. An increase in the generation capacity within FCS from renewable with balanced energy storage will increase the profit.

2.4.3   [B3] Manage Standards

There is a number of standards that are related to FCS. The standards will support energy management, reliability, safety, and mobility. Understanding standards will support the design and operation of all functions within FCS. List of related standards is shown in Table 2.4.

2.4.4   [B4] Manage FCS Risks

FCS design should have protection features to overcome different risks. Risk is estimated as likelihood multiplied by the magnitude of consequence of hazard scenarios. Hazard scenarios are identified based on a number of initiating events and escalation scenarios that reach potential consequences and losses. Table 2.5 shows potential hazard scenarios and potential protection measures.

2.4.5   [B5] Manage FCS Facility Operation

FCS facility will include building, energy, mechanical, electrical, communication, security, and fire systems. The operation of FCS will include daily activities of charging, supply, facility management, maintenance, and inspection and thermal, water, and waste management. The daily operation will ensure that profit, sales, and

Table 2.4 Standards related to charging infrastructures

Standard	Details
SAE J1772	It is an SAE standard related to IEC 62196 Type 1 and known as a J plug. It is a North American standard for electrical connectors for electric vehicles.
SAE J3068	It is an SAE standard related to electrical connectors and a control protocol for electric vehicles.
J3105	It is an SAE standard for automated connection devices (ACD) that mate chargers with battery-electric buses and heavy-duty vehicles.
IEC 62196	It is an IEC standard that describes socket-outlets, vehicle connectors, and vehicle inlets.
IEC 61851	It is an IEC standard that describes electric vehicle conductive charging systems.
ISO 17409	ISO standards include electrically propelled road vehicles, conductive power transfer, and safety requirements.
ISO 18246	ISO standards include electrically propelled mopeds and motorcycles and safety requirements for conductive connection to an external electric power supply.

Table 2.5 Hazard analysis of FCS

Hazard	Scenario	Mitigation/safeguards
Grid outage	Power fault propagation from station to grid	Proper design and operation of grid protection, monitoring to validate integrity
	High load on station, led to high load on grid	Monitoring of station load, apply demand-side management to balance charging loads
Charging station outage	Power fault propagation within station	Design MEG to provide local generation capacity and energy storage to support charging loads
	Grid disturbance leading to charging station outage	Proper design and operation of grid interface with protection layer, apply reliable monitoring to validate grid integrity
Fire	Excess heat release from heat sources such as power electronics	Thermal management, monitoring, control, and interlock systems
	External source of fire	Monitor sources of fire, heat, or ignition, and design proper protection layers to isolate propagation of source of fire. Install fire mitigation systems
Uninsulated high-voltage surface	Human shock or electrocution	Maintenance and inspection strategies to detect and repair uninsulated surfaces
		Proper selection of reliable technologies

risk management are achieved. Real-time data monitoring is performed on charging, MEG, grid supply, power back to the grid, health condition of physical systems, and weather conditions. Performance measures are evaluated and monitored with risk indexes and key performance indicators for improved operation management.

2.4.6 [B6] Manage Charging Requests

Charging requests are based on incoming EVs, EBs, and ETs. In addition, a number of incoming swapped batteries with their types and SOCs will be used to manage their charging. In order to effectively manage charges in all units, it is important to identify and model important parameters, as shown in Table 2.6.

2.5   Manage Energy System in FCS

This section will discuss requirement analysis for the energy management system within FCS. The coordination between energy management of EV [40] and energy management of FCS is important to ensure an optimized charging process for both FCS and EV.

Table 2.6 Parameters to manage charging

Parameter	Parameter
Incoming vehicle charging request	Incoming EV/EB/ET Vehicle plate Vehicle model Charging mode (wireless/wired) Battery type Current SOC Target SOC Expected range KM Charging time requirements (time range) Charging level (Level 1/Level 2/Level 3)
Incoming swapped battery	Battery type Current SOC Target SOC Expected range KM Charging time requirements (time range) Charging level (Level 1/2/3)

Table 2.7 The output power level classification for DC chargers according to CHADEMO and CCS standards

Standard	Version/power class	Output power (kW)
CHAdeMO	1.0	62.5 (500 V × 125 A)
	1.2	200 (500 V × 400 A)
	2.0	400 (1 kV × 400 A)
CCS	DC5 5	(500 V × 10 A)
	DC10	10 (500 V × 20 A)
	DC20	20 (500 V × 40 A)
	FC50	50 (500 V × 100 A)
	HPC150	150 (500 V × 300 A, 920 V × 163 A)
	HPC250	250 (500 V × 500 A, 920 V × 271 A)
	HPC350	350 (500 V × 500 A, 920 V × 380 A)

2.5.1 [C1] Manage Power from Grid

The power supply from the grid is balanced based on electricity price, on-/mid-/off-peak, energy generation capacity from MEG within FCS, charging price, and charging load profile. The performance of the power from the grid should meet target requirements of power quality, bidirectional power flow, and protection schemes.

2.5.2 [C2] Manage Energy Storage

The selection of the energy storage technology should meet fast-charging station requirements [41]. The energy storage technology could be battery, ultracapacitor, or flywheel and combinations of them to meet charging/discharging time requirements, storage capacity, control requirements, and protection requirements. There are essential functions as part of energy storage management, for example, battery management system (BMS) and the ability to monitor SOC. Lifecycle cost analysis is essential to ensure a profitable charging facility.

2.5.3 [C3] Manage Energy Sources

Similarly, energy sources within MEG in FCS will support charging capacity and balance energy supply from the grid. Renewable technologies such as PV and wind will be selected to meet FCS's design and operation requirements. Technology assessment of PV and wind with sizing and lifecycle cost analysis will ensure the profitable operation of FCS. Location and placement of PV and wind will also contribute to the performance evaluation of MEG and FCS. Design and technology specifications of power electronics to integrate energy sources will play an important role in enhancing FCS's performance.

2.5.4 [C4] Manage Energy to Grid

Energy can be sent back to the grid from two options: (a) via V2G or (b) return excess energy from MEG to the grid. The concept of V2G allows energy to be sent back to the grid based on mobility profiles where charged EV/EB/ET can be used to return energy to the grid. The local generation from MEG can provide excess energy back to the grid after supplying to charging units. The power supply to the grid is managed based on electricity price, off-/mid-/on-peaks, energy generation capacity from MEG and EV/EB/ET, charging price, and charging load profile. The decision of returning power to the grid is based on power quality, price, charging load, and demand profiles. The performance of the power to the grid should meet target requirements of power quality, bidirectional power flow, and protection schemes.

2.5.5 [C5] Manage Energy to Units

Energy supply to the charging units is based on grid inputs and energy capacity from MEG. Demand-side management is used to balance energy supply to units based on mobility and charging demand profile. Electricity price for peak period should be considered to ensure profitable operation of the charging station.

2.5.6   [C6] Manage MEG

MEG includes renewable energy resources such as PV and wind. Energy storage is included in MEG to offer a balance between supply from the grid and supply to charging units. Energy storage is used to balance energy back to the grid from charging units via V2G and from excess energy from MEG. Management of MEG includes monitoring, optimization, protection, and lifecycle cost analysis.

2.6   Manage Charging in FCS

2.6.1   [D1] Manage Fast Charging

The charging station will operate in different modes. Fast charging is used when incoming vehicles require charging in a short time using Level 3 chargers. Effective management of fast-charging stations will be achieved by controlling incoming requests to utilize fast chargers, considering arrival time and required charging end time. Optimized scheduling is achieved using objective function that considers optimized wait time, maximized number of served incoming vehicles.

2.6.2   [D2] Manage Ultrafast Charging

Similar to fast charging, ultrafast charging will be managed in view of incoming requests for ultrafast charging based on demand profile, incoming EV/EB/ET, and cost analysis to ensure profitable charging station.

2.6.3   [D3] Manage Wireless Charging

Selected vehicles will require wireless charging, which will take longer time, with less efficiency. Incoming requests to utilize wireless charging will notify the charging station with requests to charge wirelessly. Cost analysis will be performed to ensure maximum profits.

2.6.4 [D4] Manage Regular Charging

Regular charging is managed similarly based on incoming requests to charge. Typical regular charging takes a longer time and reduced costs. The management scheme will consider time, power quality, cost, mobility, and grid condition.

2.6.5 [D5] Manage Charge Batteries

Possible incoming requests will demand charging batteries and distribute them to specified locations for swapping and utilization. Due to space limitations, it is important to optimize charging units to host incoming batteries.

2.6.6 [D6] Manage V2G

Some vehicles might select V2G based on SoC condition and price model in view of peak periods. Negotiation between the vehicle owner and charging station will ensure win-win price models in view of electricity price from the utility company.

2.7 Analysis of Best Practice Charging Stations

There are worldwide initiatives to promote transportation electrification. The consensus is to enable technology while progressing standards and policies to support deployment projects. The following sections will highlight some of the best practices in different regions worldwide to support charging infrastructures, including fast charging.

2.7.1 European Distribution System Operators (DSO)

The European Distribution System Operators (E.DSO) is leading efforts to promote clean sources and integrate with transportation electrification infrastructures. There are around 39 electricity distribution system operators (DSOs) in 24 countries. It also includes national associations to support the reliability of Europe's electricity supply by enabling smart grids. E.DSO targets the Energy Union to support the EU's energy plans for climate change, energy security, and community development. This supports transportation electrification and charging infrastructures across Europe. E.DSO presented a number of R&D and demonstration projects and policy development. The European Commission is seeking the raise of energy efficiency targets at the EU level. The improved energy efficiency will reduce energy consumption of final energy to 36% and primary energy consumption to 39% by 2030.

2.7.2   Next-Generation Vehicle Promotion Center: Japan

In Japan, EV penetration in the market is around 30%. Japan is planning long-term electrification by 2050. The plan is to realize well-to-wheel zero emissions. The plan is to improve vehicle technology, innovation in vehicle utilization, and global zero emissions in energy supply. Based on the Japan Automobile Manufacturers Association, Inc. (JAMA), the number of hybrid electric vehicles sold in 2017 is 31.6% of total vehicles, which is equivalent to 1.385 million vehicles [42]. The number of battery electric vehicles was 18,000 vehicles (0.41%), and the number of plug-in hybrid electric vehicles was 36,000 (0.82%). A number of fuel cell electric vehicles sold in 2017 was 849 vehicles (0.02%).

2.7.3   US Transport Electrification

The United States is progressing in the transportation electrification plan in different directions. The promotion of PHEV as personal vehicles can contribute to over 95% of the emission reductions, while the overall transportation emission is reduced by over 80% based on full battery-based electric vehicles [43].

There are a number of R&D initiatives such as micromobility, lithium-ion battery recycling and extended life, charging infrastructures with fast charging, and hydrogen vehicle charging infrastructures.

The expansion of micromobility will increase the penetration of small and lightweight vehicles that operate at low speeds, typically below 25 km/h. Micromobility technologies include e-bikes, bicycles, shared bicycles, electric skateboards, electric scooters, and electric pedal-assisted bicycles (or pedelec). Micromobility technologies can be electric or human-powered. The expansion of electric micromobility technologies will enhance transportation electrification and mobility and reduce GHG emissions.

2.7.4   Smart City Sweden

Smart City Sweden is an initiative to promote the development of transportation electrification with sustainable technology. Sweden aims to achieve net-zero greenhouse gas emissions by 2045 [44]. The planned expansion of electric vehicle deployment led to an increase in charging infrastructure. The limited range of EV requires more effective and connected charging infrastructures, including parking (e.g., residential buildings) and charging stations.

2.7.5   Electrification of Public Bus in Singapore

The impact of transportation electrification of the public bus has demonstrated improved energy efficiency, reduced cost, and enhanced environmental performance in terms of greenhouse gas emissions. The study conducted in Singapore showed positive impacts on sustainability [45]. The study showed a 43% reduction in total lifecycle cost for the selected case study of the autonomous electric minibus. The study compared electric minibuses with 12 m diesel buses. The lifecycle GHG emissions of the 6 m autonomous electric minibus were also reduced by 47% compared with the 12 m diesel bus.

2.8   Charging Technology Specifications

There are three main standards of fast-charging systems of EVs: GB/T, CHAdeMO, and CCS. Table 7.7 lists the comparison between many versions of CHAdeMO and CCS standards showing the voltage, current, and output power (kW) they can provide.

Table 2.8 shows the EV batteries in the market with voltage, power capacity, distance range, and fast-charging power. It shows that the market has not reached above 150 kW, but only Porsche Taycan as they are claiming their fast-charging system can provide charging at 350 kW and with 800v [46, 47].

Table 2.9 lists the Fast Charger Manufacturers details, including the country; the output power (kW); the topology, that is, modular or nonmodular; the module power; the output voltage; and the cooling methods.

Table 2.8 The most common EVs and charging systems in the market

Model battery	Capacity (KWh)	Battery voltage (V)	Distance range (km)	Fast-charging power (kW)
Tesla Model S	100	350	539	120
Tesla Model X	100	350	475	120
Chevy Bolt	60	N/A	383	N/A
Renault Zoe	41	400	281	43
Nissan Leaf	40	350	270	50
Hyundai IONIQ	28	360	200	100
KIA Soul EV	30	375	174	50–120
Audi E-tron	95	400	400	150
Jaguar I-PACE	90	390	470	100
Rapide E a	65	800	N/A	100
Porsche Taycan a	95	800	500	350

Table 2.9 The manufacturers of fast-charging systems in the market

Ref	Manufacturer	Country	Output power (kW)	Modular	Module power (kW)	Output voltage (V)	Cooling
[48]	Tesla	USA	120	Yes	10	N/A	N/A
[49, 50]	Tritium	Australia	50	yes	N/A	N/A	Liquid cooling
			175–475	yes	N/A	N/A	Liquid cooling
[51]	Everyone	France	Up to 350	Yes	N/A	N/A	Air cooling
[52]	ABB	Switzerland	175–460	Yes	N/A	N/A	N/A
[53]	Chargepoint	United States	Up to 500	Yes	31.25	200–100	Liquid cooling
[54]	EVteQ	India	200	Yes	10	750	Air cooling
[55, 56]	Siemens	Germany	150	Yes	30,120	200–920	N/A
[57, 58]	PRE	Netherlands	315	Yes	AC/DC 12.5 DC/DC 25	300–850	Air cooling
[59]	PHIHONG	Taiwan	Up to 160	Yes	30	N/A	Air cooling
[60]	IES-Synergy	France	Up to 200	Yes	50	300–750	N/A

2.9 Analysis of Mobility Requirements

Mobility is defined as the ability of human or mobile entities to communicate and access services, regardless of any location changes or technical access environment [61]. Transportation electrification is based on mobility requirements, defined based on state ranges, distances, locations, mobility reasons, and timing. The mobility will support community development, and transportation electrification could play an important role to support mobility. There are several mobility categories, which include seamless mobility, nomadic mobility, terminal mobility, network mobility, personal mobility, service mobility, macro-mobility, micromobility, intra-access network mobility, inter-access network mobility, inter-core network mobility, horizontal mobility, vertical mobility, mobility in infrastructure-based networks, and mobility in non-infrastructure-based networks. Mobility requirements are reflected in number of parameters, such as number of trips, number of persons per trip, number of vehicles, number of vehicles per road, number of persons per vehicle, and number of persons per road or region. These parameters are measured per hour/day/week/month/year.

2.10   Automotive Cybersecurity

Security plays a vital role in transportation automation. One important aspect is to link vehicles with charging stations and control centers. Transportation electrification is based on real-time connections with charging infrastructures to enable timely charging based on EV battery conditions. Figure 2.3 shows the proposed cybersecurity (CS) model as part of transportation electrification. Cybersecurity algorithms and components are required to protect EV, station, and other digital infrastructures such as traffic signal, edge server, and cloud server. Cybersecurity should include sensors in vehicle, charging stations, roads, traffic, and pedestrian devices, including persons with disability. Cameras are also controlled and connected with two-way communication with proper cybersecurity controls. Safety protocols are needed for in-vehicle network (IVN) security threats.

The analysis of cyberattacks is performed using scenarios that will examine the protection of each component with different input signals. Cybersecurity is associated with different communication layers: V2V, V2I, I2V, and V2X. Cellular and dedicated short-range communication (DSRC) protocols are utilized. These communication signals are either as a direct connection or via a communication network. The cybersecurity and controls should work in different and extreme weather conditions. Cybersecurity will be integrated with collision avoidance for

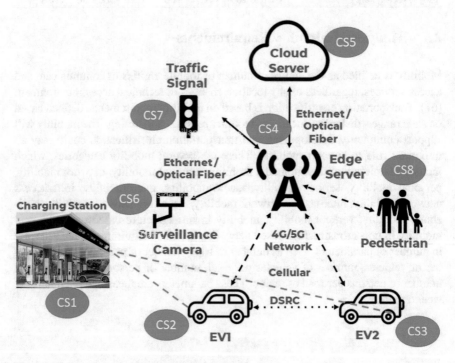

Fig. 2.3 Cybersecurity for transportation electrification

transportation and road safety. It is expected to issue emergency alert messages from cyberattacks to prevent accidents and risks to transportation electrification and traffic management. The cybersecurity process includes identifying, protecting, detecting, responding, and recovering activities. Threat modeling methods are required to ensure full control over cybersecurity for the transportation electrification infrastructure. Hazard analysis and risk assessment (HARA) and threat analysis and risk assessment (TARA) are used to ensure risk mitigation of security threats and complete protection and mitigation. The management of security threats includes preventative defense, active defense, and passive defense. Techniques are implemented for cybersecurity, including authentication, firewall, digital signature and certification, cryptography, malware detection, and intrusion detection. Blockchain is used for security in a vehicular ad hoc network (VANET). Security will be established for in-vehicle communications among devices, electronics, sensors, electronic control units (ECUs), vehicle components (display, braking, light, etc.), and passengers. All devices with the passenger are also protected with security control. The security controls are implemented for connected and autonomous vehicles and supporting systems, such as advanced driver-assistance systems (ADAS). Functions within ADAS should have proper security controls, which cover adaptive cruise control (ACC), automatic emergency braking (AEB), blind spot monitoring (BSM), forward collision warning (FCW), lane departure warning (LDW), and lane-keeping assist (LKA). There are a number of interactions between functional safety and cybersecurity and linked with cyber-physical systems (CPS). There are a number of related standards such as ISO 27001, ISO 15408, ISO 214534, and SAE J3061.

The communications between vehicle and charging station or service provider will require solid security protection layers in place. Lacking timely access to the charging station might cause threats and risks to passengers and roads.

2.11   Summary

This chapter discussed different requirements to establish a fast-charging station. The literature review section elaborated on background development and research efforts to identify requirements to support a fast-charging station design. The requirements are classified into four major categories. The first group discussed requirements related to FCS design, which covers power systems, electronics, control, protection, layout, and modeling and simulation. The second group discussed FCS facility, which includes incoming vehicles, risks, operation, and financial and change requests. The third group covered energy management, which includes power from grid, power to grid, power to charging units, energy resources, energy storage, and energy management of MEG within FCS. The fourth and last group discussed the management of charging unit, which includes fast charging, ultrafast charging, regular charging, wireless charging, charge swapped battery, and V2G. Best practices from Europe, Japan, and the United States are discussed to reflect directions and experiences in establishing fast-charging infrastructures. Mobility requirements are discussed with the considerations of fast-charging infrastructures. Requirements for cybersecurity systems, models, and operations are discussed in view of standards.

References

1. Z. Gao et al., Battery capacity and recharging needs for electric buses in city transit service. Energy **122**, 588–600 (2017). https://doi.org/10.1016/j.energy.2017.01.101
2. O. Elma, H.A. Gabbar, Sizing analysis of hybrid DC fast-charging system for electric vehicles, in *2018 International Conference on Smart Energy Systems and Technologies, SEST 2018 – Proceedings*, 2018, pp. 1–5. https://doi.org/10.1109/SEST.2018.8495688
3. SAE, Electric Vehicle Power Transfer System Using Conductive Automated Connection Devices, 2020 [Online]. Available: https://www.sae.org/standards/content/j3105_202001/. Accessed 20 Jan 2020.
4. A. Kunith, R. Mendelevitch, D. Goehlich, Electrification of a city bus network – An optimization model for cost-effective placing of charging infrastructure and battery sizing of fast-charging electric bus systems. Int. J. Sustain. Transp. **11**(10), 707–720 (2017). https://doi.org/1 0.1080/15568318.2017.1310962
5. M. Janovec, M. Koháni, Exact approach to the electric bus fleet scheduling. Transp. Res. Procedia **40**, 1380–1387 (2019). https://doi.org/10.1016/j.trpro.2019.07.191
6. Y. He, Z. Song, Z. Liu, Fast-charging station deployment for battery-electric bus systems considering electricity demand charges, Sustain. Cities Soc. **48**, February, 2019. https://doi.org/10.1016/j.scs.2019.101530
7. Y. Wang, Y. Huang, J. Xu, N. Barclay, Optimal recharging scheduling for urban electric buses: A case study in Davis. Transp. Res. Part E Logist. Transp. Rev. **100**, 115–132 (2017). https://doi.org/10.1016/j.tre.2017.01.001
8. R. Shi, J. Zhang, H. Su, Z. Liang, K.Y. Lee, An Economic Penalty Scheme for Optimal Parking Lot Utilization with EV Charging Requirements, Energies (Basel), 2020-11-01, **13**(6155), 6155
9. An examination of electric vehicle technology, infrastructure requirements and market developments. Canadian Electronic Library (Firm), distributor.; Newfoundland and Labrador. Department of Environment and Conservation. 2016. Online: https://www-deslibris-ca.uproxy.library.dc-uoit.ca/ID/10049411. Accessed 25 Jan 2022
10. Y. Zhang, J. Chen, L. Cai,J. Pan, Expanding EV Charging Networks Considering Transportation Pattern and Power Supply Limit, IEEE transactions on smart grid, 2019-11, **10**(6), pp. 6332–6342
11. S.-G. Yoon, Y.-J. Choi, J.-K. Park, S. Bahk, Stackelberg-Game-Based Demand Response for At-Home Electric Vehicle Charging. IEEE transactions on vehicular technology **65**(6), 4172–4184 (2016)
12. A. Guerrero, J. David, B. Bhattarai, R. Shrestha, T.L. Acker, R. Castro, Integrating electric vehicles into power system operation production cost models. World Electr. Veh. J. **12**(4), 263 (2021)
13. CHAdeMO Standard, 2020. [Online]. Available: https://www.chademo.com/tag/next-generation-ultra-high-power/. Accessed 28 Jan 2020
14. OppCharge [Online]. Available: https://www.oppcharge.org/. Accessed 15 Feb 2020
15. SAE, [Online]. Available: https://www.sae.org/standards/. Accessed 24 Jan 2022.
16. S. Kalker, B. Mortimer, B. Schäfer, I. Schoeneberger, M. Stieneker, W.R. Tarnate, S. Wolff, R.W. De Doncker, R. Madlener, A. Monti, D. U. Sauer, Fast-Charging Technologies, Topologies and Standards, **10**(1). RWTH Aachen University

17. Lajunen, Lifecycle costs and charging requirements of electric buses with different charging methods. J. Clean. Prod. **172**, 56–67 (2018). https://doi.org/10.1016/j.jclepro.2017.10.066
18. Z. Chen, Y. Yin, Z. Song, A cost-competitiveness analysis of charging infrastructure for electric bus operations. Transp. Res. Part C Emerg. Technol. **93**(June), 351–366 (2018). https://doi.org/10.1016/j.trc.2018.06.006
19. S. Haghbin, S. Lundmark, M. Alakula, O. Carlson, An isolated high-power integrated charger in electrified-vehicle applications. IEEE Trans. Veh. Technol. **60**(9), 4115–4126 (2011)
20. M. Budhia, G.A. Covic, J.T. Boys, C.Y. Huang Development and evaluation of single sided flux couplers for contactless electric vehicle charging, in *2011 IEEE energy conversion congress and exposition*. Sept. 2011, pp. 614–621.
21. B. Esteban, M. Sid-Ahmed, N.C. Kar, A comparative study of power supply architectures in wireless EV charging systems. IEEE Trans Power Electr **30**(11), 6408–6422 (2015)
22. S. Lukic, Z. Pantic, Cutting the cord: Static and dynamic inductive wireless charging of electric vehicles. IEEE Electrif. Mag. **1**(1), 57–64 (2013)
23. Technology improvement pathways to cost-effective vehicle electrification O.C. Onar, J.M. Miller, S.L. Campbell, C. Coomer, C.P. White, and L.E. Seiber, A novel wireless power transfer for in-motion EV/PHEV charging, in *2013 Twenty-Eighth Annual IEEE Applied Power Electronics Conference and Exposition (APEC)*, 17–21 March 2013 2013, pp. 3073–3080
24. H. Fahad, H.A. Gabbar, Wireless Flywheel-Based Fast Charging Station (WFFCS), in *2018 International Conference on Renewable Energy and Power Engineering (REPE)*, 24–26 Nov. 2018 2018, pp. 1–6.
25. Manoj Kumar M Tirupati, An Overview on Electric Vehicle Charging Infrastructure. TATA ELXSI. Accessed 1 Feb 2021 [Online]. Avilable: https://www.tataelxsi.com/Perspectives/WhitePapers/TataElxsi_Whitepaper_An_Overview_on_Electric_Vehicle_Charging_Infrastructure
26. Drive Team, Five Innovations Herald Easier and Faster EV Charging Future, September 22, 2019. Accessed 1 Feb 2021 [Online]. Available: https://driivz.com/blog/ev-charging-technology-innovations/
27. M. Yilmaz, P.T. Krein, Review of charging power levels and infrastructure for plug-in electric and hybrid vehicles, in *2012 IEEE International Electric Vehicle Conference*, 4–8 March 2012 2012, pp. 1–8.
28. N. Sakr, D. Sadarnac, A. Gaucher, A review of onboard integrated chargers for electric vehicles, in *2014 16th European Conference on Power Electronics and Applications, Lappeenranta, Finland*, 2014, pp. 1–10
29. T. Stamati, P. Bauer, On-road charging of electric vehicles, 2013 IEEE Transportation Electrification Conference and Expo (ITEC), Detroit, MI, USA, 2013, pp. 1–8. https://doi.org/10.1109/ITEC.2013.6573511
30. S.K. Nayak, Electric Vehicle Charging Topologies, Control Schemes for Smart City Application, in *2019 IEEE Transportation Electrification Conference (ITEC-India), Bengaluru, India*, 2019, pp. 1–6
31. Zidan H.A. Gabbar, Chapter 8 - Design and control of V2G, in *Smart Energy Grid Engineering*, H. A. Gabbar Ed.: Elsevier, Paperback ISBN: 9780128053430, 2017, pp. 187–205
32. K.A. Cheow, Charging Electric Vehicle, Jul 1, 2020 [Online]. Available: https://medium.com/@ackhor/charging-electric-vehicle-bc16883051dc
33. Un-Noor, Fuad; Padmanaban, Sanjeevikumar; Mihet-Popa, Lucian; Mollah, Mohammad; Hossain, Eklas, A Comprehensive Study of Key Electric Vehicle (EV) Components, Technologies, Challenges, Impacts, and Future Direction of Development, Energies (Basel), 2017-08-17, Vol.10 (8), p.1217.
34. Abul Masrur, M, Hybrid and Electric Vehicle (HEV/EV) Technologies for Off-Road Applications, Proceedings of the IEEE, 2021-06, Vol.109 (6), p.1077-1093.
35. Mahmoudzadeh Andwari, A. Pesiridis, S. Rajoo, R. Martinez-Botas, and V. Esfahanian, A review of Battery Electric Vehicle technology and readiness levels, Renewable and Sustainable Energy Reviews, vol. 78, pp. 414-430, 2017/10/01/ 2017.

36. European Automobile Manufacturers Association, A review of Battery technologies for auto-motive application. Accessed 16 Feb 2021 [Online]. Available: https://scholar.uwindsor.ca/cgi/viewcontent.cgi?article=8885&context=etd

37. Chitradeep Sen, Performance analysis of batteries used in electric and hybrid electric vehicles, Masters thesis, UWindsor, Department of Electrical and Computer Engineering,Windsor,2010. Accessed 16 Feb 2021 [Online]

38. S. Chakraborty, H.-N. Vu, M.M. Hasan, D.-D. Tran, M.E. Baghdadi, O. Hegazy, DC-DC con-verter topologies for electric vehicles, plug-in hybrid electric vehicles and fast charging sta-tions: State of the art and future trends. Energies **12**(8) (2019)

39. K.K. Gokul, B.P. Emmanuel, S. Ashok, S. Kumaravel, Design and control of non-isolated bidirectional DC-DC converter for energy storage application, in *2017 2nd IEEE International Conference on Recent Trends in Electronics, Information & Communication Technology (RTEICT), Bangalore, India*, 2017, pp. 289–293

40. Electric Vehicle Energy Management System. Accessed 18 Jan 2022. CSA-RR_ ElectricVehicle_WebRes.pdf

41. Performance analysis of batteries used in electric and hybrid-electric vehicles electric vehi-cle, Available:https://scholar.uwindsor.ca/cgi/viewcontent.cgi?article=8885&context=etd. Accessed 25 Jan 2022

42. Next Generation Vehicle Promotion Center, Japan, Accession on 24-Jan-2022, http://www.cev-pc.or.jp/english/ev-strategy.html

43. DOE and DOT Launch Joint Effort to Build Out Nationwide Electric Vehicle Charging Network. Accessed 24 Jan 2022. https://www.energy.gov/articles/doe-and-dot-launch-joint-effort-build-out-nationwide-electric-vehicle-charging-network.

44. Smart City Sweden, Online, https://smartcitysweden.com/focus-areas/mobility/electrifica-tion/. Accessed 25 Jan 2022

45. A. Pathak, G. Sethuraman, A. Ongel, M. Lienkamp, Impacts of electrification & automa-tion of public bus transportation on sustainability—A case study in Singapore. Forschung im Ingenieurwesen **85**(2), 431–442 (2020)

46. Peter Weidenhammer, A sports car that covers over 300 miles with superb performance—but without a drop of gasoline? Welcome to the future: The Mission E electric concept car – Porshe cars North America 2019 [Online]. Available: https://www.porsche.com/usa/aboutporsche/christophorusmagazine/archive/374/articleoverview/article01/. Accessed 2 Feb 2022

47. J. Winterhagen, "Mission E"– Dr. Ing. h.c. F. Porsche AG 2018 [Online]. Available: https://newsroom.porsche.com/en/products/porsche-taycan-mission-e-drive-unit-battery-charging-electro-mobilitydossier-portscar-production-christophorus-387-15827.html.   Accessed   2 Feb 2022

48. M. Alatalo, Module size investigation on fast chargers for BEV, Swedish Electromobility Centre, 2018 [Online]. Available: http://emobilitycentre.se/wp-content/uploads/2018/08/Module-sizeinvestigation-on-fast-chargers-for-BEV-final.pdf. Accessed 22 Dec 2018

49. Tritium, VEEFIL-RT 50KW DC FAST CHARGER, Tritium Pty Ltd 2019 [Online]. Available: https://www.tritium.com.au/product/productitem?url=veefil-rt-50kwdc-fast-charger. Accessed 2 Feb 2022

50. Tritium, LAUNCHING SOON- VEEFIL-PK DC ULTRA-FAST CHARGER 175 kW- 475 kW, Tritium Pty Ltd 2019 [Online]. Available: https://www.tritium.com.au/veefillpk. Accessed 2 Feb 2022

51. EVBOX, LAUNCHING SOON- VEEFIL-PK DC ULTRA-FAST CHARGER 175 kW- 475 kW, EvBox 2019 [Online]. Available: https://www.evbox.fr/produits/borne-recharge-rapide. Accessed 2 Feb 2022

52. ABB, High Power Charging, ABB 2019 [Online]. Available: https://new.abb.com/ev-charging/products/car-charging/high-powercharging. Accessed 2 Feb 2022

53. ChargePoint "ChargePoint Express Plus" ChargePoint 2019 [Online]. Available: https://www.chargepoint.com/products/commercial/express-plus. Accessed 22 Feb 2022

54. EVteQ "SET-QM EV Charging Module"Evteq 2017 [Online]. Available http://www.evteq-global.com/product-details/set-qm-evcharging-module. Accessed 2 Feb 2022
55. Siemens "High power charging for current and future ecars" Siemens Mobility2018 [Online]. Available: https://www.siemens.com/global/en/home/products/mobility/road-solutions/electromobility/ecars-high-power-charginginfrastructure.html. Accessed 2 Feb 2022
56. Siemens "Sinamics DCP" Siemens AG 2019 [Online]. Available: https://w3.siemens.com/drives/global/en/converter/dc-drives/dc-cconverters/pages/sinamics-dcp.aspx. Accessed 2 Feb 2022
57. PRE "12.5 kW Charger Module" PRE 2018 [Online]. Available: http://www.prelectronics.nl/media/documenten//12k5W_Charger_datasheet_V11.180417.194409.pdf. Accessed 2 Feb 2022
58. PRE "25 kW DC/DC Charger Module" PRE, 2018 [Online]. Available: http://www.prelectronics.nl/media/documenten//25kW_Charger_datasheet_V13.180417.194545.pdf. Accessed 2 Feb 2022
59. Phihong "Phihong EV Chargers" Phihong 2017 [Online]. Available: https://www.phihong.com.tw/newcatalog/2017%20EV%20Chargers%20%E5%86%8A%E5%AD%90_20170420_EN.pdf. Accessed 2 Feb 2022
60. Ies "Charger Module" IES Synergy [Online]. Available: http://www.ies-synergy.com/en/products/keywatt-charging-stationspower-modules/charger-module. Accessed 2 Feb 2022
61. S. Chen, Y. Shi, B. Hu, M. Ai, Mobility Management: Principle, Technology and Applications, 2016

Chapter 3
Fast-Charging Station Design

Hossam A. Gabbar, Abdalrahman Shora, Abu Bakar Siddique, and Yasser Elsayed

Acronyms

CS	Charging station
FCS	Fast-charging station
AC	Alternating current
DC	Direct current
EV	Electric vehicle
EMS	Energy management system
FESS	Flywheel-based energy storage system
PI	Proportional-integral
PMSM	Permanent magnet synchronous machine
MPC	Model predictive control
SoC	State of charge
NG	Natural gas
RNG	Renewable natural gas

The original version of this chapter was revised. The correction to this chapter is available at https://doi.org/10.1007/978-3-031-09500-9_18

H. A. Gabbar (✉)
Faculty of Energy Systems and Nuclear Science, Ontario Tech University (UOIT), Oshawa, ON, Canada

Faculty of Engineering and Applied Science, Ontario Tech University, Oshawa, ON, Canada
e-mail: Hossam.gabbar@uoit.ca

A. Shora · A. B. Siddique
Faculty of Engineering and Applied Science, Ontario Tech University, Oshawa, ON, Canada

Y. Elsayed
Faculty of Energy Systems and Nuclear Science, Ontario Tech University (UOIT), Oshawa, ON, Canada

© Springer Nature Switzerland AG 2022, Corrected Publication 2023   35
H. A. Gabbar, *Fast Charging and Resilient Transportation Infrastructures in Smart Cities*, https://doi.org/10.1007/978-3-031-09500-9_3

LNG	Liquefied natural gas
PV	Photovoltaic
V2G	Vehicle to grid
NR	Nuclear reactor
MRAC	Model reference adaptive control
MPPT	Maximum power point tracking
P&O	Perturb and observe

Nomenclature

V_{DC}	DC link voltage
V_{BT}	Battery voltage
V_{SC}	Supercapacitor voltage
D	Duty ratio
$J(\theta)$	Cost function
γ	Adaptive gain

3.1   Introduction

Transportation electrification plays an important role to achieve clean transportation. The fuel economy is improved by introducing high-performance fuel cell hybrid electric vehicle [1]. The effectiveness of the charging of electric and fuel cell vehicles is essential to achieve optimized operation of charging stations and the integration with power sources of a PV/battery/hydrogen [2]. The research on optimal operation of charging stations provided reliable strategies for charging-discharging-storage integrated stations [3]. The research has been extended to demonstrate hydrogen supply and production with sustainability considerations [4]. Lifecycle assessment of electric and fuel cell vehicles is used to evaluate different energy supply strategies, including biomass, which added value to clean charging stations [5]. The assessment of renewable natural gas refueling stations for heavy-duty vehicles is used to define gaps and opportunities for clean transportation and hybrid charging stations [6]. Hydrogen supply solutions are also assessed for hydrogen fueling stations, which offered a complementary solution to support transportation electrification [7]. There are needs to integrate renewable energy sources with charging stations to reduce lifecycle cost and GHG emissions. Wind turbines are integrated with hydrogen fueling stations with on-site hydrogen production [8]. The integration of waste-to-energy conversion and waste transportation with transportation electrifications is also studied in different regions such as island communities [9]. The integrated supply chain system with energy, transportation, and waste disposal costs is also analyzed to support charging stations [10]. The integration of heat sources is also important and can be integrated microgrids with different renewable and storage systems, which increase energy supply in charging stations [11]. Collaborative power distribution networks are studied to reduce transportation greenhouse gas emissions [12].

Charging stations and ultrafast-charging stations are presented to support mobility with economic benefits. The detailed design of charging and ultrafast stations can be improved using enhanced power electronics. Examples are shown in literature to reflect advances in improved designs of power electronic components. For example, three-phase LCL filter is presented with optimization model for enhanced electric vehicle ultrafast-charging [13]. Modular multiple-input bidirectional DC-DC power converter (MIPC) is developed for HEV/FCV applications [14].

A fast-charging station has a significant load impact on the grid. Hence, scientists are working hard to develop renewable energy-based hybrid storage systems to support the load of the station. To establish such a system, an efficient control algorithm is required to coordinate the energy production and consumption by the system. The control strategy that controls the energy exchange within the facilities is called the energy management system (EMS). With the help of this EMS, it is possible to achieve the most optimum performance from the system.

In [15], a coordinate charging strategy has been adopted to improve the demand curve of the FCS. With this control strategy, the authors could enhance the profit margin by reducing the generation cost. In [16], an intelligent charging control scheme has been adopted to reduce the demand of the load to the grid during peak times. The strategy considers the grid electricity price and the time of use to make the decision.

An energy storage system (EMS) can help to improve the effects of discontinuous charging of a Level 3 charging station. Flywheel is an excellent medium for energy storage and can be utilized extensively within fast-charging facilities. For a flywheel-based energy storage system (FESS), controlling its speed in the desired value is essential while keeping the DC bus voltage constant. In [17], a proportional-integral (PI) controller has been used to regulate DC bus voltage and speed of the flywheel. In this work, a permanent magnet DC machine was used to emulate the operating behavior of a flywheel and then coupled with a permanent magnet synchronous machine (PMSM) to extract the DC link voltage.

In [18], the author proposed an adaptive fuzzy logic controller for a hybrid fuel cell/battery system in an EV. With this control algorithm, effective power allocation and real-time control were achieved. Moreover, this strategy adopted the power loss model to enhance the charging efficiency by solving the problem of optimal control and charging control.

In many works of literature, model predictive control (MPC) has been used as an energy management system of hybrid energy systems that combines energy storage systems with renewables. In [19], an energy management system based on an MPC control system has been developed and implemented within the ultracapacitor to control its state of charge (SoC). However, the author did not consider the system cost while considering a short-time operation. The optimal scheduling online distributed model predictive control scheme has been proposed in [20] for electric vehicle charging stations. This EMS scheme considered the voltage and power flow constraints to reduce the energy cost of a charging station. An intelligent fuel cell/battery hybrid vehicle has been proposed in [21] integrated with powertrain modeling and motion control. In this study, a nonlinear MPC-based EMS system has been modeled to solve the problem related to integrated control. But the authors neglected the device cost and their replacement cost.

The fast-charging stations also rely heavily on the power converters especially the DC/DC converters to connect the energy storage devices and input power sources to the DC link that will be connected to the electric vehicle's battery to charge it. A hybrid energy system is preferred in fast-charging stations to combine the high energy density from a device like the battery and high power density from a device like a supercapacitor or flywheel. This hybrid system can be utilized also to mitigate the load from the grid, as charging one EV is equal to the load of one house. The energy storage devices can be charged during the off-peak hours and discharged during the peak hours, and this could also reduce the cost of the charging. To control and transfer the power between the DC link and the energy storage devices, the bidirectional DC/DC converters are used. In addition, the energy storage devices can transfer the power between them like transferring the power between the supercapacitor and the battery and vice versa with these converters, and this could be useful to increase the battery's lifetime as its state of charge (SoC) should not be more than 80% and not less than 20% [22]. To improve the efficiency of the system and increase the power rate of charging, it is important to design DC/DC converters with high efficiency and low power losses. The battery electric vehicles (BEVs) and plug-in hybrid electric vehicles (PHEVs) utilize different energy storage devices with different power and energy characteristics; therefore, several studies have been conducted to utilize different topologies of DC/DC converters with these energy storage devices [23]. Using multi-input converters has several merits compared to single-input DC/DC converters, such as minimizing the component numbers, minimizing the power conversion stages, and centralizing the control, which makes them a good selection for numerous applications such as plug-in electric vehicles and hybrid electric vehicles as the load can be supplied from different input voltages to various output levels. Many studies in multi-input converters have been conducted to obtain different topologies as well as to combine one or more of its merits and to add the feature of the modularity to the converter to make integrating more power sources to the converter more flexible. Despite combining all the features and the advantages in one topology is not possible. The structure of multi-input converters is divided into isolated and non-isolated converters similar to the conventional converters. The isolated converters have more size and weight compared to the non-isolated converters. Using the multi-input DC-DC converters in fast-charging stations can reduce the parts and cost of the charging and the losses of the charging as well as centralize the control. The multi-input DC/DC converters have different topologies, and each topology has its features and characteristics such as the operation modes, the number of power devices, the power flow capability, the number of input sources, and the modularity. With low part numbers and continuous input current, the multi-input converter in [24] is used in renewable energy applications, but it has unidirectional power flow. The load can be supplied with different voltages from two sources without circulating current using the multi-input converter in [25], but it has a lack of modularity. The proposed converter in [26] has fewer conduction losses and can supply the load individually or simultaneously from two different sources that have different voltage-current characteristics with three power switches only; therefore, it is used in DC microgrid applications, but the sources cannot transfer the power between them.

3.2   Conceptual Design of Fast-Charging Models

3.2.1   Functional Modeling of Fast-Charging Station

The engineering design of fast-charging station (FCS) is based on requirement analysis that covers charging station requirements, maritime demand profiles, and load profiles for local waterfront infrastructures. Functional modeling is used to identify the main functions of the target FCS to support the detailed design process. Figure 3.1 shows the functional modeling of FCS, which includes three main functions: energy management, charging, and facility management.

3.2.2   Fast Charging from the Grid

The conceptual design of fast-charging station includes power electronics of filter connected to the grid, with bidirectional AC-DC converter and bidirectional DC-DC converter connected to the load EV, which could be a battery or ultracapacitor, as shown in Fig. 3.2.

3.2.3   Fast Charging from Grid with Flywheel and Battery

Flywheel is a potential energy source for fast charging where it is based on kinetic energy, with high reliability and short response time for charging and discharging.

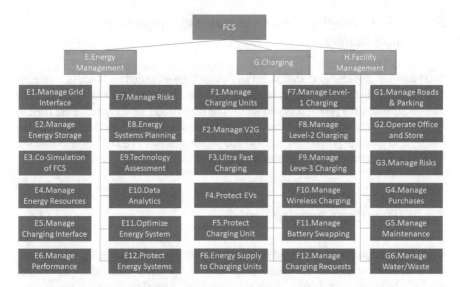

Fig. 3.1 Functional model of fast-charging station (FCS)

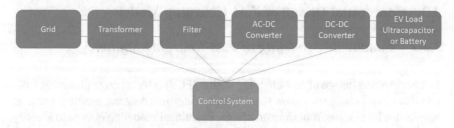

Fig. 3.2 Conceptual design of fast-charging station

The conceptual design of fast-charging station with grid connection, flywheel, and battery system is shown in Fig. 3.3. Bidirectional DC-DC converter is used to establish the connection with the battery system.

3.2.4 Fast Charging with Micro Energy Grid

Micro energy grid is integrated within the fast-charging station to expand the energy generation capacity with reduced load on the grid and in case of gird outage. The conceptual design of the fast-charging station with micro energy grid is shown in Fig. 3.4. AC bus is used to connect flywheel, wind, and the grid with the DC bus via AC-DC converter and controller. Battery, solar, and fuel cell are connected to the DC bus and can be linked to the AC bus to support other AC loads. Bidirectional DC-DC converter is connecting the micro energy grid with the EV charging loads.

3.2.5 Fast Charging from Grid with Supercapacitor and Battery

The proposed fast-charging station in Fig. 3.5 uses the proposed multi-input converter to charge and discharge simultaneously or individually the energy storage devices as well to exchange the power between them. The battery and supercapacitor are used to form a hybrid energy storage system, to combine the high energy density from the battery and the high power density from the supercapacitor. The DC bus is connected to DC/DC buck converter to control the charging rate of the electric vehicle.

3.2.6 Powering Charging Station

Charging station can be powered based on its location. Ship powering can be gas-based or electric-based. Gas-based ships can be operated with diesel, hydrogen, biofuel, NG/RNG, or LNG. Electric-based ship powering is based on nuclear

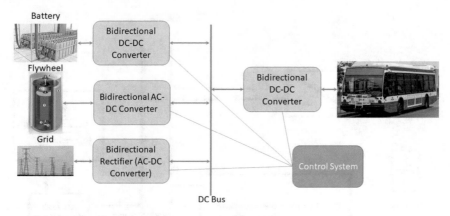

Fig. 3.3 Fast charging from grid, flywheel, and battery system

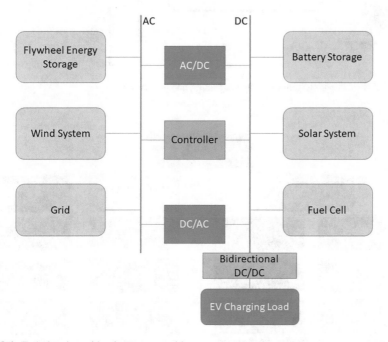

Fig. 3.4 Fast charging with micro energy grid

reactor, battery, ultracapacitor, PV, or fuel cell. Powering onshore charging station (CS) can be achieved by the grid, nuclear, PV, wind, hydrokinetics, or fuel cell. Powering offshore CS can be via nuclear, PV, wind, hydrokinetics, fuel cell, or hydropower. It is possible to use combinations of these options for powering ships, onshore, or offshore CS. Figure 3.6 shows the possible options for powering charging stations and maritime ships.

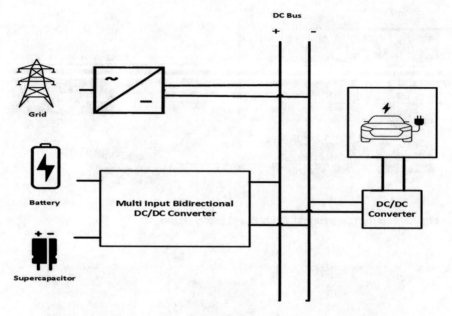

Fig. 3.5 Fast charging from grid, supercapacitor, and battery system

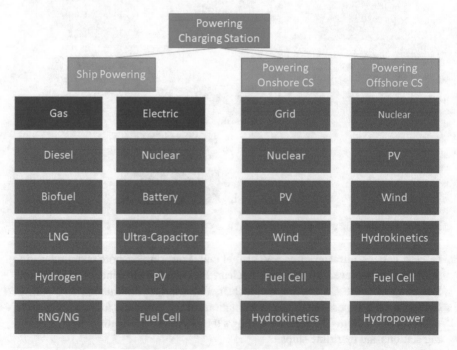

Fig. 3.6 Powering charging stations

3.2.7   FCS Cyber Physical System Modeling

The physical system model of FCS includes three main subsystems: facility, charging units, and energy systems. The detailed components of each subsystem are described in Fig. 3.7.

3.2.8   Physical System Modeling for Maritime and Charging Station

In order to complete the detailed design of the hybrid energy system for maritime system, possible design components are defined for both the ship and charging station, as shown in Fig. 3.8.

3.3   Detailed Design of Fast-Charging Station

This section will discuss detailed design of fast and ultrafast-charging stations in view of detailed requirements presented in the previous chapter.

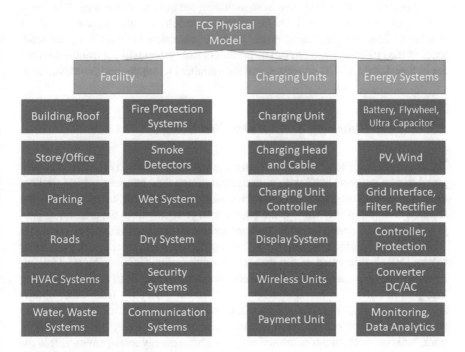

Fig. 3.7   FCS physical model

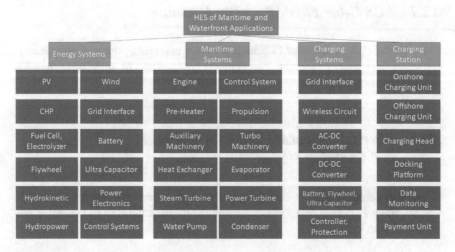

Fig. 3.8 Physical system model for HES with charging station for maritime systems

3.3.1 Fast-Charging Station Design

The fast-charging schematic is shown in Fig. 3.9, where it includes the DC charging station side and the EV side. SBEA (snowball endurance algorithm) is used to offer battery management system functions, including control and monitor of EV battery. SBEA will communicate with the fast-charging station controller which will offer a number of functions, including power control, protection, optimization, and resiliency of the fast-charging station. The power control system will control the rectifier, the bidirectional DC-DC converter, and the variable DC supply and protection unit.

3.3.2 Fast-Charging Station Detailed Design

The detailed design schematic of fast-charging station is shown in Fig. 3.10. The proposed design supports V2G with bidirectional power flow between vehicle, station, and the grid. The integration of renewables with the station is important to satisfy the required energy capacity while balancing the energy supply from the grid. PV is integrated via DC-DC converter, while the wind turbine is integrated via AC/DC converter.

In the case of large charging loads for electric buses and electric trucks, nuclear reactor (NR) is integrated within the fast-charging station (FCS) using load following mechanisms to ensure adaptive energy generation from NR to meet the charging load profiles, with the considerations of generation capacities from other renewables and grid condition. The design will consider different control strategies based on optimization objectives. Energy management is established at each vehicle, truck, and bus, with proper communications with the energy management within the

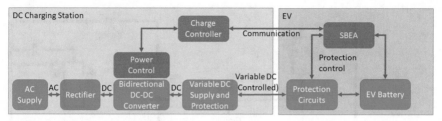

Fig. 3.9 Fast-charging station design

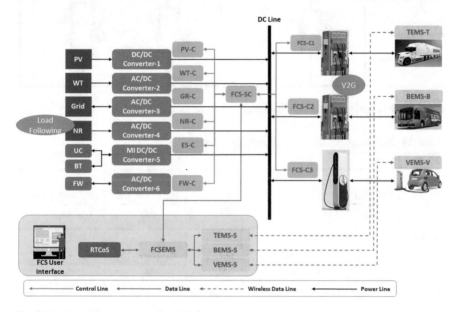

Fig. 3.10 Fast-charging station detailed design

station. The option for real time co-simulation will support the evaluation of different operation scenarios prior to the application of changes.

3.3.3 Detailed Design of Multi-Input Converter for Fast-Charging Station

The multi-input consists of six metal–oxide–semiconductor field-effect transistor (MOSFET), two inductors, and one capacitor as shown in Fig. 3.11, and the design can be modular by adding two power switches and one inductor to combine the switching leg. Table 3.1 shows the equations used to calculate the minimum values of inductors L_1 and L_2, and Table 3.2 shows the design specifications of the proposed converter.

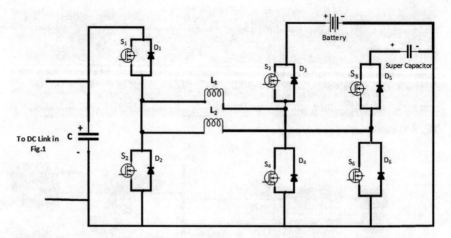

Fig. 3.11 Schematic of multi-input converter

Table 3.1 Inductance equations for different operation modes

Mode	1	2	3	4	5
L_1, min	$\dfrac{(1-D).T_s}{\Delta I_{L1}}.V_{DC}$	–	$\dfrac{D.T_s}{2\Delta I_L}.V_{Bt}$	$\dfrac{V_{DC}.(1-d_2).T_s}{\Delta I_L}$	$\dfrac{(V_{DC}-V_{Bt})d_1.T_s}{\Delta I_{L1}}$
L_2, min	–	$\dfrac{(1-D).T_s}{\Delta I_{L2}}.V_{DC}$	$\dfrac{D.T_s}{2\Delta I_L}.V_{Bt}$	$\dfrac{V_{DC}.(1-d_2).T_s}{\Delta I_L}$	$\dfrac{(V_{DC}-V_{SC})d_1.T_s}{\Delta I_{L2}}$

Table 3.2 Design specifications of the proposed converter

Specification	Battery voltage (V_{BT})	Supercapacitor voltage (V_{SC})	DC link voltage (V_{DC})	Switching frequency (f_s)	Inductors (L_1 and L_2)	Capacitor (C)	Power
Values	200 V	160 V	500 V	20 kHz	0.75 mH and 0.75 mH	500 µF	15 W

3.3.4 The Operation Modes of the Converter

The proposed converter works in five different operation modes as we will discuss later to transfer the power between the DC link and energy storage devices simultaneously or individually and to exchange the power between them. The five operating modes of the converter are the following:

3.3.4.1   Mode 1: Battery to DC Link

The inductor L_1 is charged and discharged during this operation mode in three time intervals T_1, T_2, and T_3 to discharge the battery only in the DC link to charge the electric vehicle's (EV's) battery. Under the steady-state condition, the relation between DC link voltage as an output and battery voltage as input is expressed using Eq. (3.1):

$$V_{DC} = \frac{T_1}{T_2 + T_3}.V_{BT} = \frac{D}{1-D}.V_{BT} \tag{3.1}$$

where D is the duty cycle ratio defined by $\frac{T_1}{T}$ and T_S is the total period of the switching cycle. The battery voltage is boosted to the DC link with working in duty cycle equal to $D > 0.5$. The DC link charge the battery, by reversing the current in L_1. Under the steady-state condition, and with taking into consideration D to be T_1/T_S, the relation between DC link voltage and battery voltage can be expressed using Eq. (3.2):

$$V_{BT} = \frac{T_1}{T_2 + T_3}.V_{DC} = \frac{D}{1-D}.V_{DC} \tag{3.2}$$

3.3.4.2   Mode 2: Supercapacitor to DC Link

This operating mode is used to utilize the merit of the high power density of the supercapacitor to charge the EV in less time. The inductor L_2 is charged and discharged in the DC link in three time intervals T_1, T_2, and T_3 similar to the previous operation mode. Eq. (3.3) shows the voltage of the DC link V_{DC} as an output voltage as a function of the supercapacitor V_{SC} input voltage:

$$V_{DC} = \frac{T_1}{T_2 + T_3}.V_{SC} = \frac{D}{1-D}.V_{SC} \tag{3.3}$$

where D is the duty cycle ratio equal to $\frac{T_1}{T_S}$. To charge the supercapacitor from the DC link, the current flow reversed in L_2 and by operating at $D < 0.5$. Equation (3.4) expresses the relation between input and output voltage considering duty ratio D to be T_1/T_S:

$$V_{SC} = \frac{T_1}{T_2 + T_3}.V_{DC} = \frac{D}{1-D}.V_{DC} \tag{3.4}$$

3.3.4.3   Mode 3: Battery and Supercapacitor

In mode 3(a), the supercapacitor can be charged from the battery. The converter boosts the battery voltage V_{BT} to charge the supercapacitor as shown in Eq. (3.5) using the switches S_3, S_5, and S_6:

$$V_{SC} = \frac{T_s}{T_2 + T_3}.V_{BT} = \frac{1}{1-D}.V_{BT} \tag{3.5}$$

The lifetime of the battery can be increased by charging it from the supercapacitor using mode 3(b). The switches S_3, S_5, and S_6 are turned on to reverse the current through the inductors, as the battery can be charged in the buck operation mode using Eq. (3.6):

$$V_{BT} = \frac{T_1}{T_s}.V_{SC} = D.V_{SC} \tag{3.6}$$

The switching sequence of the power switches in the three time intervals of the previous operation modes is summarized in Table 3.3.

3.3.4.4   Mode 4: Battery and Supercapacitor to DC Link

In this operation mode, the load from the grid can be mitigated during the peak hours, by charging the EV from the battery and the supercapacitor simultaneously, as the EVs consume high power and increase the load on the grid. Inductors L_1 and L_2 are charged and discharged in five time intervals that are represented in Table 3.4 to supply the DC link simultaneously from the battery and the supercapacitor using Eqs. (3.7) and (3.8):

$$V_{BT} = \frac{T_4 + T_5}{T_1}.V_{DC} = \frac{T_s - d_2 T_s}{d_1 T_s}.V_{DC} = \frac{1-d_2}{d_1}.V_{DC} \tag{3.7}$$

$$V_{SC} = \frac{T_4 + T_5}{T_{1+} T_{2+} T_3}.V_{DC} = \frac{T_s - d_2 T_s}{d_2 T_s}.V_{DC} = \frac{1-d_2}{d_2}.V_{DC} \tag{3.8}$$

where d_1 is the ratio of the on-time of switch S_3 to total switching period T_S and, similarly, d_2 corresponds to switch S_2.

Table 3.3 Modes 1, 2, and 3 switching states

	Mode 1(a)	Mode 1(b)	Mode 2(a)	Mode 2(b)	Mode 3(a)	Mode 3(b)
T_1	S_2, S_3	S_1, S_4	S_2, S_5	S_1, S_6	S_3, S_6	S_3, S_5
T_2	D_4, D_1	D_2, D_3	D_6, D_1	D_2, D_5	S_3, D_5	D_6, S_3
T_3	S_1, S_4	S_2, S_3	S_6, S_1	S_2, S_5	S_3, S_5	S_6, S_3

3.3.4.5   Mode 5: DC Link to Battery and Supercapacitor

In contrast to operation mode 4, during the off-peak hours, the DC link can charge the energy storage devices simultaneously to save the cost of energy. Using Eqs. (3.9) and (3.10) and in the same way in mode 4, inductors L_1 and L_2 will be charged and discharged in five time intervals as represented in Table 3.4 to charge the supercapacitor and battery:

$$V_{BT} = \frac{T_1 + T_5}{T_{1+}T_{2+}T_{3+}T_5}.V_{DC} = \frac{T_1 + T_5}{T_s - T_4}.V_{DC} = \frac{d_1}{d_2}.V_{DC} \tag{3.9}$$

$$V_{SC} = \frac{T_1 + T_5}{T_s}.V_{DC} = d_1.V_{DC} \tag{3.10}$$

3.4   Control System Design

The proposed control system design is based on defined control strategies for each controller and subsystem as well as for the integrated charging station. The detailed illustration of the control strategy is presented in Fig. 3.12.

Table 3.4 Switching states in modes (4) and (5)

	T_1	T_2	T_3	T_4	T_5
Mode 4	S_2, S_3, S_5	S_2, D_4, S_5	S_2, S_4, S_5	D_1, S_4, D_6	S_1, S_4, S_6
Mode 5	S_1, S_3, S_5	D_2, S_3, S_5	S_2, S_3, S_5	S_2, S_4, S_5	S_1, D_3, S_5

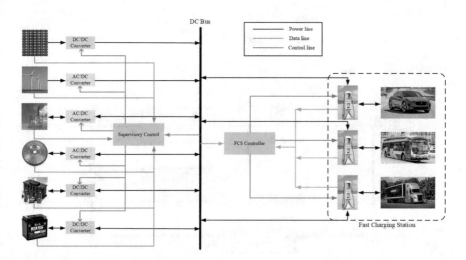

Fig. 3.12 Illustration of the detailed design of the control strategy

3.4.1 Control Design for Charging Unit

3.4.1.1 Model Reference Adaptive Control

The design of control strategy options for each charging unit is illustrated in Fig. 3.13. In this study, a model reference adaptive control (MRAC) strategy has been adopted for each of those units. This control technique includes four different blocks, namely, process, control unit, reference model, and adaptive mechanism. Each of these blocks has different functions. Reference model is used to determine the desired response where the control unit helps to provide the necessary control signal to achieve the actual performance from the process. The adaptive mechanism provides adaptation parameter θ to the control unit for high tracking performance. Several techniques like Lyapunov theory, augmented error theory, and MIT law are available to design the adaptive mechanism. In this study, MIT law is used to do this.

The tracking error (e) can be obtained from the difference between process response and reference model response as follows:

$$e = Y_P - Y_M \tag{3.11}$$

Based on this error, a cost function ($J(\theta)$) can be determined by the following formula:

$$J(\theta) = \frac{e^2(\theta)}{2} \tag{3.12}$$

Here, θ is an adaptive parameter and is considered as the proportion of negative gradient of J so that the cost function becomes zero:

$$\frac{d\theta}{dt} = -\gamma \frac{\delta J}{d\theta} = -\gamma e \frac{\delta e}{\delta\theta} sign(e) \tag{3.13}$$

where

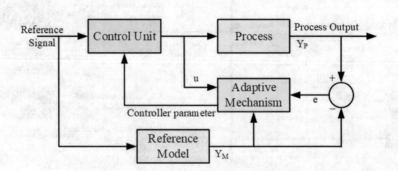

Fig. 3.13 Control algorithm of model reference adaptive control

$$sign(e) = \begin{cases} 1, e > 0 \\ 0, e = 0 \\ -1, e < 0 \end{cases} \tag{3.14}$$

The process transfer function is considered as $qG(s)$ where q is an unknown parameter and $G(s)$ is a second-order transfer function. Transfer function for the reference model is taken as $q_0G(s)$, where the known parameter is represented by q_0:

$$e = Y_P - Y_M$$

$$e = qG\theta U - q_0 GU \tag{3.15}$$

The updated rule is

$$\frac{\delta e}{\delta \theta} = qGU = \frac{q}{q_0} Y_M \tag{3.16}$$

From (3.13), the updated MIT rule is

$$\frac{d\theta}{dt} = -\gamma' \frac{q}{q_0} Y_M e = -\gamma Y_M e \tag{3.17}$$

Here, γ signifies the adaptive gain and is considered as positive small amount.

3.4.1.2   Maximum Power Point Tracking (MPPT)

MPPT or maximum power point tracking is a strategy to extract the maximum power from the variable power sources. The technique is commonly used in the solar system and turbine system. Several control strategies like perturbed and observed (P&O) method, incremental conductance, current sweep, constant voltage, and temperature method can be utilized for MPPT, but perturbed and observed method is the most commonly used control technique among them. In this P&O method, the controller adopts hill climbing method to adjust the voltage from the source and measures power. This method continues to adjust power in the direction of increment or decrement. It is a hill climbing method since it depends on the power rising curve against voltage below the maximum power point and fall above that point.

3.4.2 Integrated Control Design for the Charging Station

When each of those power components are designed, the overall system is then integrated with each other with their control strategy within the fast-charging station. The detailed design of the integrated control design is illustrated in Fig. 3.14. Here, MRAC controller can be used for utility grid control, fast-charging station control, battery, supercapacitor, and flywheel control whereas MPPT control strategy can be used for the renewable sources like PV and wind turbine system.

3.4.3 Energy Management System (EMS)

After designing the control strategy, an efficient energy management system is required within the system for the optimum operation of the overall system. This EMS system will work as a supervisory controller which will collect the

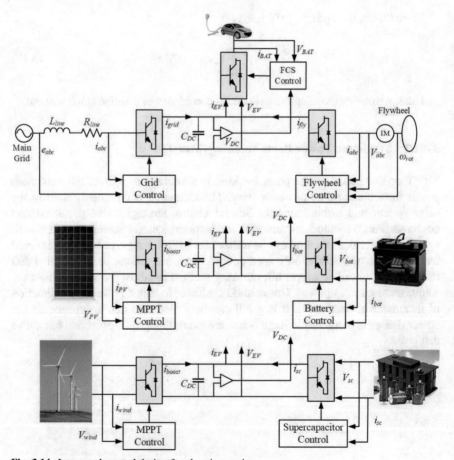

Fig. 3.14 Integrated control design for charging station

information from all of those components, and based on the load demand, power generation capacity of each of those power sources, and power generation price, the system will decide which system to adopt to meet the power demand and hence optimize the total cost and generation of the complete system. The schematic of an energy management system for the fast-charging station is given in Fig. 3.15.

3.5 Summary

This chapter presented the design of fast-charging station. The design included conceptual design, functional modeling, physical system model, and designed design. The proposed design includes hybrid energy storage, renewable energy resources, and the interface to the grid. Design of power electronics is presented where multi-input converter is illustrated with different operation scenarios for charging and discharging from different energy sources. Control strategies are discussed using integrated control system. Energy management schemes are presented to support the operation of the fast-charging station.

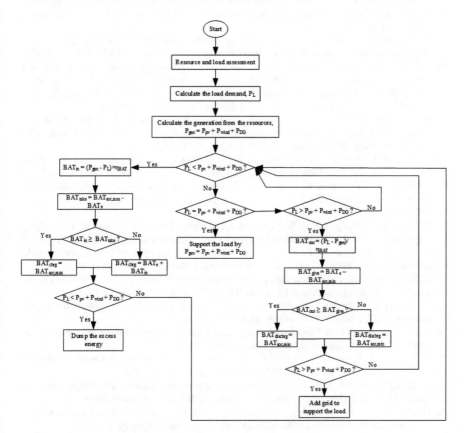

Fig. 3.15 Energy management system for fast-charging station

References

1. S. Ahmadi, S.M.T. Bathaee, A.H. Hosseinpour, Improving fuel economy and performance of a fuel-cell hybrid electric vehicle (fuel-cell, battery, and ultra-capacitor) using optimized energy management strategy. Energy Convers. Manag. **160**, 74–84 (2018)
2. P. García-Triviño, J.P. Torreglosa, F. Jurado, L.M. Fernández Ramírez, Optimised operation of power sources of a PV/battery/hydrogen-powered hybrid charging station for electric and fuel cell vehicles. IET Renew. Power Gener. **13**(16), 3022–3032 (2019)
3. P. Yang, X. Yu, G. Liu, Research on optimal operation strategy of fuel cell vehicle charging-discharging-storage integrated station. IOP Conf. Ser. Earth Environ. Sci. **675**(1), 12139 (2021)
4. N. Norouzi, Hydrogen production in the light of sustainability: A comparative study on the hydrogen production technologies using the sustainability index assessment method. Nucl. Eng. Technol. (2021)
5. B. Singh, G. Guest, R.M. Bright, A.H. Strømman, Life cycle assessment of electric and fuel cell vehicle transport based on forest biomass: LCA of EV and FCV using energy from biomass. J. Ind. Ecol. **18**(2), 176–186 (2014)
6. W. Yaïci, M. Longo, Assessment of renewable natural gas refueling stations for heavy-duty vehicles. J. Energy Resour. Technol. **144**(7) (2022)
7. P. Song, Y. Sui, T. Shan, J. Hou, X. Wang, Assessment of hydrogen supply solutions for hydrogen fueling station: A Shanghai case study. Int. J. Hydrog. Energy **45**(58), 32884–32898 (2020)
8. X. Wu, H. Li, X. Wang, W. Zhao, Cooperative operation for wind turbines and hydrogen fueling stations with on-site hydrogen production. IEEE Trans. Sustain. Energy **11**(4), 2775–2789 (2020)
9. Z. Zsigraiová, G. Tavares, V. Semiao, C.M. de Graça, Integrated waste-to-energy conversion and waste transportation within island communities. Energy (Oxford) **34**(5), 623–635 (2009)
10. P. Hasanov, M.Y. Jaber, S. Zanoni, L.E. Zavanella, Closed-loop supply chain system with energy, transportation and waste disposal costs. Int. J. Sustain. Eng. **6**(4), 352–358 (2013)
11. A.R. Jordehi, Scheduling heat and power microgrids with storage systems, photovoltaic, wind, geothermal power units and solar heaters. J. Energy Storage **41**, 102996 (2021)
12. N. Danloup, V. Mirzabeiki, H. Allaoui, G. Goncalves, D. Julien, C. Mena, A. Hamid, A. Choudhary, Reducing transportation greenhouse gas emissions with collaborative distribution: A case study. Manag. Res. Rev. **38**(10), 1049–1067 (2015)
13. D. Cittanti, F. Mandrile, M. Gregorio, R. Bojoi, Design space optimization of a three-phase LCL filter for electric vehicle ultra-fast battery charging. Energies (Basel) **14**(5), 1303 (2021)
14. A. Hintz, U.R. Prasanna, K. Rajashekara, Novel modular multiple-input bidirectional DC–DC power converter (MIPC) for HEV/FCV application. IEEE Trans. Ind. Electron. **62**(5), 3163–3172 (2015)
15. Z. Xu, Z. Hu, Y. Song, Z. Luo, K. Zhan, J. Wu, Coordinated charging strategy for PEVs charging stations, in *2012 IEEE Power and Energy Society General Meeting*, (IEEE, 2012), pp. 1–8
16. Y. Cao, S. Tang, C. Li, P. Zhang, Y. Tan, Z. Zhang, J. Li, An optimized EV charging model considering TOU price and SOC curve. IEEE Trans. Smart Grid **3**, 388–393 (2011)
17. X. Zhang, J. Yang, A DC-link voltage fast control strategy for high-speed PMSM/G in flywheel energy storage system. IEEE Trans. Ind. Appl. **54**(2), 1671–1679 (2017)
18. J. Chen et al., Adaptive fuzzy logic control of fuel-cell-battery hybrid systems for electric vehicles. IEEE Trans. Industr. Inform. **14.1**, 292–300 (2016)
19. Z. Jia et al., A real-time MPC-based energy management of hybrid energy storage system in urban rail vehicles. Energy Procedia **152**, 526–531 (2018)
20. Y. Zheng et al., Online distributed MPC-based optimal scheduling for EV charging stations in distribution systems. IEEE Trans. Industr. Inform. **15**(2), 638–649 (2018)
21. H. Zheng et al., Integrated motion and powertrain predictive control of intelligent fuel cell/battery hybrid vehicles. IEEE Trans. Industr. Inform. **16**(5), 3397–3406 (2019)
22. Government of Canada, Available online https://open.canada.ca/data/en/dataset/1920f1ba-14ba-48c1-ba8c-05dbfeed75e6. Accessed 14 June 2020

23. Clean Energy Canada, Available online https://cleanenergycanada.org/canada-targets-100-zero-emission-vehicle-sales-by-2040/.Accessed 14 June 2020
24. E.D. Kostopoulos, G.C. Spyropoulos, J.K. Kaldellis, Real-world study for the optimal charging of electric vehicles. Energy Rep. **6**, 418–426 (2020). https://doi.org/10.1016/j.egyr.2019.12.008
25. S. Chakraborty, H.-N. Vu, M.M. Hasan, D.-D. Tran, M.E. Baghdadi, O. Hegazy, DC-DC converter topologies for electric vehicles, plug-in hybrid electric vehicles and fast charging stations: State of the art and future trends. Energies **12**, 1569 (2019). https://doi.org/10.3390/en12081569
26. M. Dhananjaya, S. Pattnaik, Design and implementation of a multi-input single-output DC-DC converter, in *2019 IEEE International Conference on Sustainable Energy Technologies and Systems (ICSETS)*, (2019), pp. 194–199

Chapter 4
Analysis of Transportation Electrification and Fast Charging

Yasser Elsayeda, Hossam A. Gabbar, Otavio Lopes Alves Estevesa, Ajibola Adelekea, and Abdalrahman Elshorab

Acronyms

AC	Alternating current
AVTA	Antelope valley transit authority
BEB	Battery electric bus
BMS	Battery management system
CNG	Compressed natural gas
CO_2	Carbon dioxide
CTA	Chicago Transit Authority
DC	Direct current
DCFC	Direct current fast charging
EB	Electric bus
EDG	Equivalent diesel gallon
ET	Electric truck
EV	Electric vehicle

The original version of this chapter was revised. The correction to this chapter is available at https://doi.org/10.1007/978-3-031-09500-9_18

Y. Elsayeda · O. L. A. Estevesa · A. Adelekea
Faculty of Energy Systems and Nuclear Science, Ontario Tech University (UOIT), Oshawa, ON, Canada

H. A. Gabbar (✉)
Faculty of Energy Systems and Nuclear Science, Ontario Tech University (UOIT), Oshawa, ON, Canada

Faculty of Engineering and Applied Science, Ontario Tech University, Oshawa, ON, Canada
e-mail: Hossam.gabbar@uoit.ca

A. Elshorab
Faculty of Engineering and Applied Science, Ontario Tech University, Oshawa, ON, Canada

EVSE   Electric vehicle supply equipment
FCS   Fast-charging station
FCV   Fuel cell vehicle
GHG   Greenhouse gas
GVWR   Gross vehicle weight rating
HC   Hydrocarbons
HDT   Heavy-duty trucks
HP   Horsepower
HVDC   High-voltage direct current
kW   Kilowatt
kWH   Kilowatt hour
MEG   Micro energy grid
MT   Megatons
Nox   Nitrogen oxides
OBC   On-board charging
OEM   Original equipment manufacturer
PM   Particulate matter
SoC   State of charge
USA   United States

4.1   Introduction

Recently, relevant changes in the global power sector have been driven by a confluence of factors. In addition to a growing incentive to foment renewable sources in the global energy mix, motivated by the decarbonization of the energy system as part of efforts to climate change mitigation, there is also a surge in demand for advanced technology, smart cities, and electric mobility. However, several elements have become a concern for electric utilities, which include the asynchrony between generation and consumption, the insertion of a diverse range of loads into the grid, and the grid resilience against power outages.

One sector that has been expected to have a significantly increased demand in the near future is electric vehicles (EVs); however, this scenario also brings some concern for the electric sector, and one of these problems is that the local power grid is not available everywhere. In long-distance trips, several stops for charging would be necessary, and it is often impracticable to connect to the utility grid, so it is necessary to create isolated systems along the way. In addition, since the use of electric cars is basically to decrease carbon emissions, it is also important to ensure that the charging is being made from sustainable and clean energy sources.

In Canada, the transportation sector is the second largest source of GHG emission as it is reported as 25% of the total GHG emission. From 1990 to 2018, there was an increase in emissions with an average growth rate of 0.7% annually. It has increased from 614 megatons (MT) of CO_2 to reach 729 in 2018. Researchers have identified the breakdown of the transportation GHG emission between different subsectors to understand potential strategies better to mitigate [1].

The gradual change from traditional vehicles to electric vehicles has many environmental and economic advantages. However, as the number of electric vehicles increases, so does the need for charging [2]. With the increasing interest in reducing carbon emission worldwide as a means of mitigating climate change, the adoption of EVs will bring a sustainable future for the next generation of automobiles. In recent years, the market penetration of EVs has increased significantly. However, the integration of EVs into the grid poses more challenges for engineers around the world. The evaluation of potential grid impact due to EV integration is essential to guarantee optimal grid operation. A number of studies on power system impact cover several aspects of the problem [3].

Charging stations for electric vehicles are basically an energy system infrastructure that has the purpose of supplying energy to the vehicles for charging the battery bank. It could be divided depending on which type of energy network is used, direct current (DC) and alternating current (AC).

The chapter is organized as follows: Sect. 4.2 will survey the charging systems for e-bus in the literature, by presenting the fast-charging challenges and opportunities and making an introduction of the technologies and topologies of the fast-charging system for e-bus; Sect. 4.2 also presents the market of e-buses comparing the fuel-based, hybrid, and electric buses. Section 4.3 introduces an environmental and economic comparison between the fuel-based trucks and hybrid electric, plug-in hybrid electric, and battery electric trucks. Section 4.3 also shows the structure of the charging system of electric trucks and presents the scenario of e-trucks in the market. Section 4.4 will introduce the main differences between the charging station technologies, pointing out the advantages and drawbacks of each topography. And to sum up, Sect. 4.5 resumes the present work.

4.2 Analysis of Electric Buses

The recent Canadian statistics show a dozen of thousands of buses that support mass transit [4]. According to the statistics, there are 45.8 million passenger trips on the transit network of Canada's urban electric buses. Buses have produced a steady increase in GHG emissions over the period.

On the other hand, the buses can transfer 40–80 people as a capacity. It is environmental and economically feasible to increase the electrification of electric buses. This chapter lists and analyzes the main challenges of the fast-charging system for e-bus. In addition, the fast-charging processes and technologies include a converter, energy storage systems, control methods, and power electronic devices. Also, the market will be explored to show the current e-bus charging systems available in the market, including their pros and cons.

4.2.1   e-Bus Opportunities

There are many opportunities for e-bus over conventional diesel buses like the lower emission of GHG as the e-bus has zero GHG emissions, including CO_2, NOx, HC, and PM. Also, the fuel cost is due to the high efficiency of the battery electric bus (BEB) than the diesel bus. The BEB uses about 2.17 kWh per mile or 17.35 miles per an equivalent to a diesel gallon which is four times more efficient than the diesel bus, which can route for only 4.2 miles for the same amount of equivalent diesel gallon (EDG) [5]. In addition, the maintenance cost of the e-bus is lower than the conventional fuel-based bases because the number of movable parts is reduced in the e-bus. The average maintenance cost for one mile is $0.64 for e-bus while $0.88 for the conventional buses [6]. Also, the e-bus is more comfortable than the combustion-based buses because the earlier is based on magnetic induction and hence less vibration [7].

4.2.2   e-Bus Challenges

There are a couple of challenges for deploying e-bus. The first is the economic barrier as the conventional bus is lower than the BEB for the same capacity as the diesel bus costs about $445,000. In contrast, the equivalent bus with the same capacity of passengers costs $770.000 besides the charging station infrastructure and operation [8]. Also, the planning for the BEB is complicated as it depends on many factors, such as the route and the size of the e-bus, and needs comprehensive training for the drivers to optimally operate the BEB. In addition, the battery is sufficient for a specific amount of travelling distance, and increasing the battery increases the weight. On the other hand, it is a developing technology that risks the absence of support or spare parts.

4.2.3   Battery Technologies

Lithium-ion batteries have a higher power and energy density. Therefore, they are most preferable for e-bus [9]. The batteries of the bus are embedded with the bust structure on the back, top, or bottom according to OEM (original equipment manufacturer) [10]. The battery bank is sized based on the equivalent horsepower and the equivalent tank of diesel. For example, a battery in BEB of 1 kW is equivalent to 1.34 horsepower (HP), and battery storage of 105 kWh is equivalent to 2.8 gallons of diesel [11].

4.2.3.1   Battery Size and Range

The battery size of e-bus depends on the charging technology used. In general, they are in the range of 76 kWh like Nova Bus LFSe and up to 660 kWh like Proterra Catalyst E2 [12]. The range depends on the capacity, environment, and route variables.

4.2.3.2 Battery Aging

The battery aging depends on the charging and discharging cycles, the chemical and physical structure degradation, the heating and temperature, and the usage [13]. Also, one of the factors affecting the battery's aging is keeping the battery at high SOC. It is better to keep it from 30% to 70% than between 70% and 90% [14]. However, when the battery reaches the end of its useful lifetime, it can give up to 60% of its capacity as stationary storage in the charging station.

4.2.4 Depot for Bus Charging

The depot bus charging systems are installed where the off-duty buses are stopped. The typical charging time is less than 5 h and the power range from 40 kW to 200 kW. It is cheaper than on-route charging. Figure 4.1 shows the buses charging from depot charging [15]. It is preferred for small scale as they are cheaper than the on-route charging systems. The Chicago Transit Authority (CTA) is one example of a BEB fleet that has been in service since 2014 in Chicago City in the United States. Two buses charge in 3–5 h using a 100 kW charging system that charges 300 kWh for each pack [16]. They work for 6 continuous hours in the morning, and then they are charged during midday for 3–5 h. After charging, they operate for the afternoon shift. So, they can operate for 80 miles through 8 h.

4.2.5 On-Route Charging

These charging systems utilize high-power and fast chargers that take a few minutes to charge fully. They are of power capacity of 200–500 kW. The batteries can be smaller in size as they can charge less time and frequently. Ten transits in the United

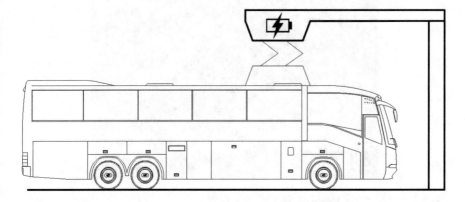

Fig. 4.1 The depot bus charging

States use on-route charging to connect to depot charging [15]. There are three charging methods in on-route charging: conductive, inductive, and battery swapping.

4.2.6 Conductive On-Route Charging

In conductive on-route charging, the buses have a pantograph that is roof-mounted to the bus, which connects to the catenary for charging. It needs space for mounting and is subject to a collision. The typical power ranges from 175 to 500 kW, and the charging process takes 5–20 min to fully charge [15].

Figure 4.2 shows a case study of a transit overhead 500 kW fast charger based on a conductive fast-charging system. There is a three-bus fleet with 15 BEB in this transit. It utilizes two on-route charging of power 500 kW, and it has 88 kWh battery packs. The charging time is 7 h to run 13.5 h a day and charge on average 13 times a day, with each recharge averaging 20 kWh.

4.2.7 Inductive On-Route Charging

There are two coils in the inductive charging method, one on board the bus and the other buried at bus stops. The capacity of this charging system is 50 kW–250 kW. The efficiency of the inductive is less than that of conductive charging and higher in cost. The problem with this method is that buses must stop and wait for charging at stops [15, 18]. Figure 4.3 shows the inductive on-route charging of three BEBs which are charging for 4 h for the full charging with two 40 kW.

Fig. 4.2 Foothill transit overhead 500 kW fast charger [17]

Fig. 4.3 Wireless
advanced vehicle
electrification charging
station for AVTA,
embedded in the street

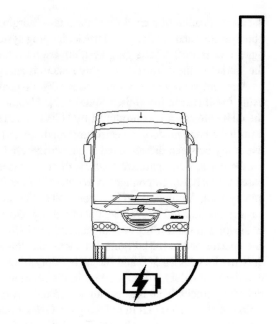

4.2.8 Battery Swapping On-Route Charging

This is the least common on-route charging method in which a fully charged battery replaces the battery of low SOC. The main issues in this method are the potential risk of damaging the battery during the replacement process and the capital cost of having a backup battery in service and others fully charged for playing as a backup battery.

The transportation sector is recognized as the largest share of GHG emissions. Electric buses in the form of transit play an essential part in transportation in Canada and the United States and can transfer millions of people daily. So electrification of buses can help immensely in reducing GHG emissions. In this chapter, the charging infrastructure and topologies are presented, and the charging methods show the pros and cons of each method with a case study and examples.

4.3 Analysis of Electric Trucks

Trucks are classified based on the loaded weight by calculating the gross vehicle weight rating (GVWR). For example, there are eight classes of e-truck from the light- to heavy-duty trucks in the United States. However, heavy-duty trucks are about a third of the total electric vehicles (EVs). They consumed a vast amount of fuel. Therefore, they emit a large amount of GHG emissions. Therefore, the electrification of e-trucks is crucial in decreasing GHG emissions. This chapter introduces the electrification of e-trucks and their fast-charging systems, showing the different opportunities and challenges in truck charging systems. In addition, different topologies and aspects of charging systems are introduced and analyzed. Furthermore, the market of the e-trucks' charging system is analyzed comparatively.

A recent survey shows that about 80% of heavy-duty trucks (HDTs) are still using fossil fuel in the United States [19]. Also, it was recorded that about 25% of the GHG emission was produced by HDTs in 2013 [20]. Many research contributions have been made for electrifying medium and HDT. For example, a comparative study between diesel-based and electric-based trucks has been conducted [21]. Furthermore, a comparative analysis of compressed natural gas (CNG) and diesel-based buses has been conducted to determine the impact on the climate and the air quality [22]. A comparison between CNG, diesel, and hybrid electric transit buses has been conducted to evaluate the efficiency and environmental performance [22]. Evaluating the feasibility of using natural gas in medium- and heavy-duty trucks has been performed in [23]. Figure 4.4 shows the classes of trucks, including the light-duty trucks Classes 3–6 and the heavy-duty trucks from Classes 7 to 8. It is clear that the heavy-duty trucks are less in distribution. They consume much more than light-duty trucks. That means they are effectively emitting more GHG and electrification of HDT is crucial in reducing the GHG emission.

In the United States, Classes 7 and 8 of HDTs can weigh up to 11.8 tons and consume 15% of the total energy of the transportation sector. HDTs also are responsible for 15% of the total GHG emission of the transportation sector [24]. In addition, the operation cost of the HDTs is very high [25]. Therefore, electrification of HDTs is very feasible to reduce the maintenance cost and the GHG emission [26].

On the other hand, surveying the literature shows the main challenges in HDT electrification: low battery densities and high capital cost [27–32]. Furthermore, another study shows that the battery bank weight is another challenge to add to the list for electrification of HDTs. Also, on-road operation of HDT in the United States weighs up to 33 tons and is concerned with time, route, and volume [33]. To sum up, surveying the recent transportation reports for the United States, it is found that more than 92% of HDT are running on fossil fuel. So, Class 8 HDTs consume about 17% of the total fuel and produce the largest share of GHG emissions. Therefore,

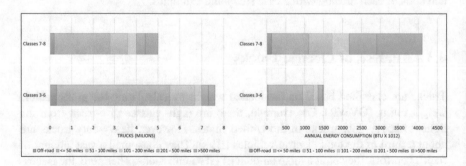

Fig. 4.4 Energy consumption of all classes of trucks

electrification of HDT is crucial for reducing pollution and is targeted by the US government immensely [23]. In this chapter, the electrification of trucks is analyzed and discussed, showing the charging systems, cost, key challenges, and opportunities.

Table 4.1 lists the components of each HDT type and the capital, operating, and maintenance cost [35]. It shows the common components commonly inherent in all truck technologies [27–32]. In addition, it lists the operating and maintenance cost of diesel, biodiesel, hybrid electric, and battery electric trucks. Figure 4.5 depicts the comparison between all types regarding the operating cost and shows the battery electric trucks. In addition, the co-emission of the HDT based on diesel and CNG is listed in Table 4.2, showing the breakdown of the CO emission for diesel and CNG gas. In addition, Fig. 4.6 shows the environmental comparison in the form of the air pollutant emission of the different types of HDT technologies, including all GHG components [23]. It shows that hybrid and battery electric HDTs are the least life-times of GHG emission while diesel and biodiesel are the highest.

4.3.1 Fast-Charging System of HDT

There are two main classes of HDT, which are Classes 7 and 8. On the scale of gross vehicle weights (GVWs), Class 7 HDTs are between 26000 and 33000 LB and have three or more axles, while Class 8 is enormous and greater than 33000 LB and is

Table 4.1 List of HDT technologies, including components and cost

Truck technology	Components	Cost	Total
Common component	Truck manufacturing	$107,362	$139,862
	Trailer manufacturing	$32,500	
Diesel	Diesel fuel production	$1,030,445	$1,255,318
	Maintenance	$224,873	
Biodiesel	Biodiesel fuel production	$867,976	$1,090,996
	Maintenance	$223,020	
CNG	Natural gas manufacturing	$855,785	$974,558
	Metal tank heavy gauge manufacturing	$60,495	
	Infrastructure	$58,278	
BE	Maintenance	$211,314	$914,410
	Power generation	$380,211	
	Battery system manufacturing	$162,000	
	Battery replacement	$160,885	
Hybrid	Maintenance	$224,873	$990,095
	Diesel fuel production	$757,262	
	Battery system manufacturing	$3000	
	Battery replacement	$4960	
	Motor	$9290	$224,393
	Power electronics	$12,388	
	Maintenance	$202,715	

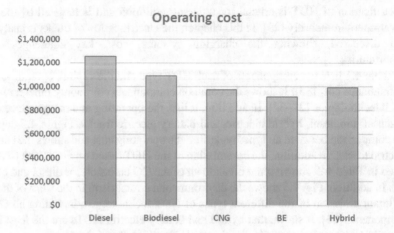

Fig. 4.5 Operating cost of different technologies of HDTs

Table 4.2 List of HDT CO emission factors for diesel and CNG fuels

Phase	North American Industry Classification System (NAICS) sector	CO emissions (t/$1 million dollars)	
		Diesel	CNG
Vehicle manufacturing and refueling infrastructure	Heavy-duty vehicle manufacturing	3.27	3.27
	Trailer manufacturing	4.5	4.5
Fuel production (based on fuel consumption)	Automotive mechanical and electrical repair and maintenance	1.38	1.38
Operation/use	Tailpipe (g/mile)	1.57	24.23

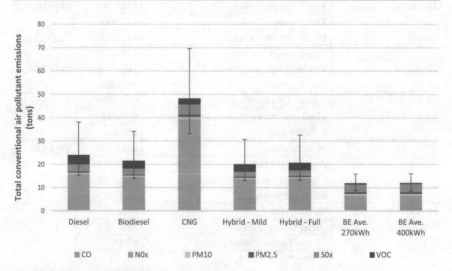

Fig. 4.6 Lifecycle air pollutant emissions of HDTs [23]

called severe duty. Garbage and sweeper trucks and bus transit are Class 7, and cement and concrete trucks are specimens of Class 8 HDTs. The following subsections show the structure and topologies of fast-charging systems of Class 7 and Class 8 HDT.

4.3.2   Depot Charging Electrical Distribution System

Figure 4.7 shows a typical electrical distribution system for the depot charging station. It shows that the high-tension voltage is stepped down from transmission lines (>110 kv) to medium voltage about 35 kv. Then it is passed to the distribution transformer to be reduced to low voltage (480v) to be used by EVSE to supply the charging station. In addition, the components, prices, and duration of the distribution system of the depot charging are listed in Table 4.3.

4.3.3   Charging Scheduling Algorithm

The charging process is a trade-off between the charging power and the charging time for heavy-duty trucks. For example, transit buses run in schedules and should be charged during off-duty periods. Therefore, the charging process should be fast

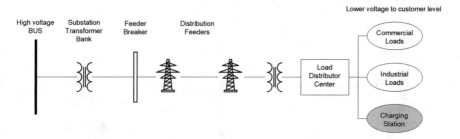

Fig. 4.7   Electrical distribution system for depot charging

Table 4.3   lectrical distribution system for depot charging

Component	Upgrade	Cost	Duration (days)
Distribution substation	Feeder breaker 5 MW	US$400,000	6–12
	Substation upgrade 10 MW	US$3–5 million	12–18
	New substation installation 10 MW	US$4–35 million	24–48
Distribution feeder	Install/upgrade feeder circuit 5 MW	US$2–12 million	3–12
Utility on site	Install distribution transformer 200 kW	US$12,000–175,000	3–8
Customer on site	350 kW DCFC EVSE	US$1200–5000	3–10

and needs high power for rapidly charging within the limited off-duty time. Figure 4.8 shows on the left part of the figure how charging with high power 100 kW is fast charging and helpful to complete charging in plenty of time during the off-duty period. On the other hand, the right part of the figure shows how charging time can be divided into two off-duty periods but using lower power.

4.3.4   Electric Truck Opportunities and Challenges

In terms of challenges, there are many barriers in HDT electrifications like range, charging times, capacity, and temperature impacts, as specified in details below.

Range: Hauling heavy loads for long distance is a big challenge and causes the battery to be drained faster than normal. Fast charging is required besides an optimization algorithm to handle such a challenge. The load weight is another issue that should be considered as the increase in load increases the electrical capacity and limits travel distance.

The only solution to cope with the range of weight and huge loads is to increase battery size. For example, in Class 8 HDT, the battery packs are designed to supply 300 kW. However, the charge time becomes a consequent result of increasing the battery. Hence, fast charger technology is required to ride the charging time issue.

By increasing the battery pack for coping with the HDT loads and distance, the weight of the batteries is added to the main load.

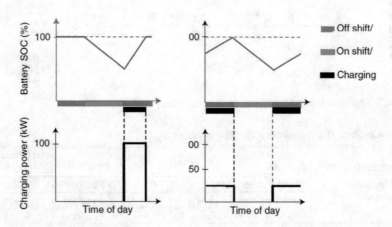

Fig. 4.8 Charging HDT power and time

4.3.5   HDT Fast Charging in the Market

The analysis of the HDT market will provide a better understanding of fast-charging demand. It is expected to see boos in the sales of medium-duty and heavy-duty e-trucks. According to the Navigant Research, the increase in penetration of e-trucks is due to stricter emission targets for commercial vehicles and government incentives for clean fuel fleets [34]. The increase in sales of e-trucks of the size of medium- and heavy-duty is estimated to grow from 31,000 vehicles in 2016 to 332,000 vehicles by 2026.

The analysis of fast charging for electric trucks (e-trucks) is analyzed. The truck technologies are classified based on weight. The analysis of different truck categories is discussed and mapped to other factors related to fast charging. Data analysis of medium-duty and heavy-duty trucks is used to understand fuel consumption and compare it with the electrification of trucks. The analysis of GHG emissions is discussed in view of charging e-trucks. Discussions about different opportunities and challenges are related to truck charging systems. Moreover, different topologies of e-truck charging systems are explained.

4.4   EV Charging Technologies

Charging stations for electric vehicles are basically an energy system infrastructure that has the purpose of supplying energy to the vehicles for charging the battery bank. It could be divided depending on which type of energy network is used, direct current (DC) and alternating current (AC). The main difference between these two systems is that in AC the battery charge is made through the vehicle's onboard controller while in DC charger, the systems charge directly the vehicle's battery. Table 4.4 lists how the DC and AC charging stations can be divided according to the power and charging capacity.

Table 4.4   Details the charging stations classified based on power level [53]

EVSE type	Power supply	Charger power	Charging time (approximate) for a 24-kWh Battery
AC charging station: L1 residential	120/230 V_{AC} and 12 A to 16 A (Single Phase)	Approximately 1.44 kW to approximately 1.92 kW	Approximately 17 h
AC charging station: L2 commercial	208–240 V_{AC} and 15 A to approximately 80 A (Single/split phase)	Approximately 3.1 kW to approximately 19.2 kW	Approximately 8 h
DC charging station: L3 fast chargers	300 to 600 V_{DC} and (Max 400 A) (Poly phase)	From 120 kW up to 240 kW	Approximately 30 Min

4.4.1   AC Charging Station

AC charging topologies are the most used models nowadays, especially due to their low-power operations and readiness to be connected to the grid utility, commonly available in commercial voltage rates (110–240 Vac), and they are subdivided into Levels 1 and 2.

Level 1 charges are single-phase converters, specially designed for residential applications, and are able to draw a current of up to 16A. The charging time for a small electric vehicle, which typically has a 24 kWh battery, is around 12–17 h to fully charge the battery bank.

On the other hand, Level 2 chargers are designed for mid-sized vehicles and usually use polyphase AC sources and are able to draw from the grid up to 80A and power level up to 20 kW. As a matter of comparison, to charge the same small vehicle with a 24 kWh battery bank, Level 2 chargers would take about 8 h to fully charge the batteries (Fig. 4.9).

4.4.1.1   Level 1 Charging

This charger typically has 2–5 miles per hour of charging. Level 1 alternating current (AC) equipment (commonly referred to simply as Level 1) charges through a 120-volt (V) AC plug. Most, if not all, PEVs will come equipped with a Level 1 cord set, eliminating the need for extra charging equipment. A regular NEMA connection (e.g., a NEMA 5–15, which is a common three-prong home plug) is on one end of the cable, and an SAE J1772 standard connector is on the other end (often referred to simply as J1772, shown in the above image). The NEMA connection fits into a conventional NEMA wall outlet, while the J1772 connector plugs into the car's J1772 charging port. It is worth noting that Tesla automobiles have a special hookup. A J1772 adaptor is included with every Tesla car, allowing it to utilize non-Tesla charging equipment. Level 1 charging is often utilized when just a 120 V outlet is available, such as while charging at home, although it may easily meet all of a driver's charging demands. For a mid-sized PEV, for example, 8 h of charging at 120 V may restore around 40 miles of electric range. In the United States, less than 5% of public EVSE ports were Level 1 as of 2020 [35] (Fig. 4.10).

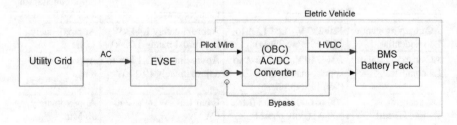

Fig. 4.9   AC charging station

Fig. 4.10 Type 1 AC
charger

4.4.1.2   Level 2 Charging

This charger typically has 10–20 miles per 1 h of charging. AC Level 2 equipment
(also known as Level 2) charges via a 240 V (common in residential applications)
or 208 V (usual in commercial applications) electrical supply. Because Level 2
equipment can charge a normal PEV battery overnight, most residences have 240 V
service, and PEV owners frequently install it for home charging. Public and work-
place charging is also often done with Level 2 equipment. This charging option has
a maximum current of 80 amperes (Amp) and a maximum power of 19.2 kW. Most
household Level 2 equipment, on the other hand, use less electricity. Many of these
machines offer 7.2 kW of electricity and run at up to 30 amps. A separate 40-Amp
circuit is required for these machines. Over 80% of public EVSE ports in the United
States were Level 2 by 2020 [35]. The J1772 connection is used by both Level 1 and
Level 2 charging devices. Level 1 and Level 2 charging devices are compatible with
all commercially available PEVs. Tesla automobiles feature a special connection
that is compatible with all of their charging alternatives, including Level 2 destina-
tion chargers and home chargers. A J1772 adaptor is included with every Tesla car,
allowing it to utilize non-Tesla charging equipment (Fig. 4.11).

4.4.2   DC Charging Station

For fast-charging stations, DC systems are the most suitable due to their capacity of
providing a very high-power level in the range of 120–240 kW and are categorized
as Level 3 chargers. DC chargers are able to fully charge the battery bank in minutes. However, these chargers are more complex because, in order to provide such a
high-power level, it is necessary to include modular converters.

These converters, when installed inside the vehicle, increase significantly the
weight and consequently the vehicle performance. To avoid this issue, these converters need to be implemented together with the charger infrastructure, and for this
reason, the charging process is made connecting the EV charge directly to the battery bank of the vehicle bypassing the onboard charger, as shown in Fig. 4.12. This
process allows the charging process to happen much faster than using the onboard
converter.

This charger typically has 60–80 miles per 20 min of charging. Rapid charging
is possible at established stations using direct current (DC) fast-charging technology (usually 208/480 V AC three-phase input). In the United States, about 15% of

Fig. 4.11 Type 2 AC
charger

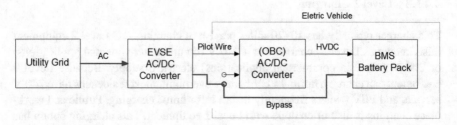

Fig. 4.12 DC charging station

public EVSE ports had DC fast charges by 2020 [35]. Depending on the kind of
charge port on the car, there are three types of DC fast-charging systems: SAE combined charging system (CCS), CHAdeMO, and Tesla. The CCS connection (also
known as the J1772 combo) is unique in that it allows a driver to charge with Level
1, Level 2, or DC fast equipment using the same charge port. The DC fast-charging
connection includes two more bottom pins, which is the only difference. The most
prevalent of the three connection types is the CHAdeMO connector. Tesla automobiles feature a special connector that works with all of their charging options,
including the supercharger, which is a rapid charging option. Although Tesla automobiles lack a CHAdeMO charging port and do not include a CHAdeMO adaptor,
the company does offer one (Figs. 4.13, 4.14 and 4.15).

4.4.3   EV Charging Standards

There are three main standards of fast-charging systems of EVs: GB/T, CHAdeMO,
and CCS. Table 4.5 lists the comparison between many versions of CHAdeMO and
CCS standards showing the voltage, current, and output power (kW) they can
provide.

Table 4.6 shows the EV batteries in the market with voltage, power capacity,
distance range, and fast-charging power. It shows that the market has not reached
above 150 kW, but only Porsche Taycan as they are claiming their fast-charging
system can provide charging at 350 kW and with 800v [36, 37].

Table 4.7 lists the Fast Charger Manufacturers details, including the country; the
output power (kW); the topology, that is, modular or nonmodular; the module
power; the output voltage; and the cooling methods.

Fig. 4.13 Combined
charging system [35]

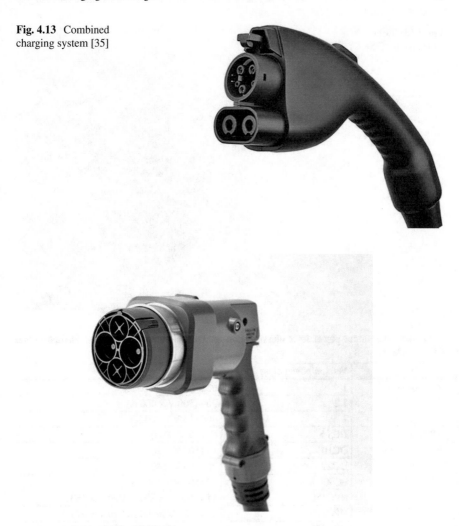

Fig. 4.14 CHAdeMO charger [35]

4.4.4 EV Fast-Charging Applications and Their Challenges

In all areas of transportation, including aircraft, trains, ships, and electric cars, there
has been a continuous shift from petroleum-based to electric-based transportation
during the last few decades (EVs). The advantages, political incentives, and drop-
ping prices, notably owing to large-scale manufacturing, are likely to accelerate this
change [51–53]. According to the US Energy Information Administration, the
world's crude oil supply will suffice until around 2050 [54]. Alternatives to fossil
fuels must clearly be found for transportation, energy generation, and other applications. EVs have long been seen as a viable transportation option, and their acceptance is growing as technology advances and economic practicality becomes a

Fig. 4.15 Tesla
supercharger [35]

Table 4.5 The output power level classification for DC chargers according to CHAdeMO and CCS standards

Standard	Version/power class	Output power (kW)
CHAdeMO	1	62.5 (500 V × 125 A)
	1.2	200 (500 V × 400 A)
	2	400 (1 kV × 400 A)
CCS	DC5 5	(500 V × 10 A)
	DC10	10 (500 V × 20 A)
	DC20	20 (500 V × 40 A)
	FC50	50 (500 V × 100 A)
	HPC150	150 (500 V × 300 A, 920 V × 163 A)
	HPC250	250 (500 V × 500 A, 920 V × 271 A)
	HPC350	350 (500 V × 500 A, 920 V × 380 A)

reality. EVs provide enhanced efficiency (energy savings) through improved fuel economy and lower emissions/pollution (particularly when power is supplied from renewable resources like wind and solar), and they assist the United States to have a broader variety of transportation fuel options. Almost the majority of the electricity generated in the United States comes from domestic sources, such as natural gas, coal, nuclear power, and renewable energy. While just 17% of US petroleum consumption was imported in 2017 [55, 56], if reserves are depleted, this may change.

Historically, it has been faster and more convenient to recharge from petroleum-based sources; therefore, charging periods have been an Achilles heel of electric transportation technology. The US Department of Energy has classified charging of electric vehicles into three levels: Level 1 is regular charging, which happens when

Table 4.6 The most common EVs and charging systems in the market

Model battery	Capacity (kWh)	Battery voltage (V)	Distance range (km)	Fast-charging power (kW)
Tesla Model S	100	350	539	120
Tesla Model X	100	350	475	120
Chevy Bolt	60	N/A	383	N/A
Renault Zoe	41	400	281	43
Nissan Leaf	40	350	270	50
Hyundai IONIQ	28	360	200	100
KIA Soul EV	30	375	174	50–120
Audi E-tron	95	400	400	150
Jaguar I-PACE	90	390	470	100
Rapide E	65	800	N/A	100
Porsche Taycan	95	800	500	350

Table 4.7 The manufacturers of fast-charging systems in the market

Refs.	Manufacturer	Country	Output power (kW)	Modular	Module power (kW)	Output voltage (V)	Cooling
[38]	Tesla	United States	120	Yes	10	N/A	N/A
[39, 40]	Tritium	Australia	50	yes	N/A	N/A	Liquid cooling
			175–475	yes	N/A	N/A	Liquid cooling
[41]	Everyone	France	Up to 350	Yes	N/A	N/A	Air cooling
[42]	ABB	Switzerland	175–460	Yes	N/A	N/A	N/A
[43]	ChargePoint	United States	Up to 500	Yes	31.25	200–100	Liquid cooling
[44]	EVteQ	India	200	Yes	10	750	Air cooling
[45, 46]	Siemens	Germany	150	Yes	30,120	200–920	N/A
[47, 48]	PRE	Netherlands	315	Yes	AC/DC 12.5 DC/DC 25	300–850	Air cooling
[49]	PHIHONG	Taiwan	Up to 160	Yes	30	N/A	Air cooling
[50]	IES Synergy	France	Up to 200	Yes	50	300–750	N/A

the charge power is less than 5 kW; Level 2 is fast charging, which occurs when the charge power is between 5 kW and 50 kW; and Level 3 is superfast charging, which occurs when the charge power is higher than 50 kW [57]. An off-board charger is used for Level 3 charging, which means the charger is located outside the car. Because of the high-power charging levels of Level 3 charging, carrying the requisite power electronics onboard is unfeasible owing to space limits. Level 3 charging often delivers electricity to the car as DC to decrease the mass of onboard electronics, but Level 1 and Level 2 charging frequently have onboard electronic converters that allow for AC energy transmission. If one concentrates on recharging an electric vehicle to enable an average daily drive of approximately 31.5 miles per day for a US driver at 10 kW charging [58], and given that typical EVs have efficiencies in the range of 240–300 Wh/mile with performance-based variations [59], this still amounts to about 1 h of charging. As a result, if the car is required again, such as for an additional nighttime journey or an emergency, alternate quicker charging alternatives should be accessible.

For example, refueling stations in North America may dispense fuel at speeds of up to 10 gallons per minute (United States) or 38 liters per minute (Canada), equating to an energy transfer rate of roughly 1.3 GJ per minute, or 21.5 MW [60, 61]. Even if only around 25% of the energy is used to power wheels (i.e., the energy in the fuel that is actually used), that still equates to a 5.4 MW effective energy transfer rate [62]. Fast charging is clearly required for customer convenience in order to avoid limiting drivers to the daily average driving distance (31.5 miles/day) and to improve the energy transfer rate to levels comparable to those seen at gasoline and diesel filling stations.

4.5　Summary

Recently, relevant changes in the global power sector have been driven by a confluence of factors. In addition to a growing incentive to foment renewable sources in the global energy mix, motivated by the decarbonization of the energy system as part of efforts to climate changes mitigate, there is also a surge in demand for advanced technology, smart cities, and electric mobility.

In addition, with all the evidence of the impacts of global warming, people have become more aware and prone to eco-friendly activities. Since transportation is indispensable for people, the importance of zero-emission vehicles has become more obvious, and for this reason, electric vehicles (EVs) are expected to spread around the world in the near future. And this tendency stands up as a promisor key to fighting climate change since transportation is the major contributor to GHG emissions.

Research shows that investments in large vehicles, such as buses and trucks, by replacing their internal combustion engines with electric motors will bring more benefits to society due to the large amount of emissions that fossil fuel-based buses and trucks produce. Researches on this topic are important to analyze the main technologies that are currently in the market, by stating the challenges and drawback of each one, and to foment invest in the electric vehicle section, especially in the development of advanced energy storage systems and technologies to decrease the charging time, which are the main concerns in this sector.

References

1. https://www.canada.ca/en/environment-climate-change/services/environmental-indicators/greenhouse-gas-emissions.html. Accessed 2 Feb 2022
2. S. Deb, K. Tammi, K. Kalita, P. Mahanta, Impact of electric vehicle charging station load on distribution network. Energies **11**(1), 178 (2018). https://doi.org/10.3390/en11010178
3. C.H. Dharmakeerthi, N. Mithulananthan, T. Saha, Impact of electric vehicle fast charging on power system voltage stability. Int. J. Electr. Power Energy Syst. **57**, 241–249 (2014)
4. Statistics Canada. Table 23-10-0067-01 Vehicle registrations, by type of vehicle
5. L. Eudy, R. Prohaska, K. Kelly, M. Post, *Foothill Transit Battery Electric Bus Demonstration Results. NREL/TP-5400-65274* (NREL, Golden, 2016) https://doi.org/10.2172/1237304
6. C. Johnson, E. Nobler, L. Eudy, M. Jeffers, *Financial Analysis of Battery Electric Transit Buses. NREL/TP-5400-74832* (NREL, Golden, CO, 2020) https://afdc.energy.gov/files/u/publication/financial_analysis_be_transit_buses.pdf
7. Learn Engineering. How Does an Electric Car Work? | Tesla Model S. May 30, 2017. https://www.youtube.com/watch?v=3SAxXUIre28
8. R. Nunno, Fact Sheet: Battery Electric Buses: Benefits Outweigh Costs. Environmental and Energy Study Institute. October 26, 2018. https://www.eesi.org/papers/view/fact-sheet-electricbuses-benefits-outweigh-costs
9. A. Perner, J. Vetter, Lithium-ion batteries for hybrid electric vehicles and battery electric vehicles, in *Advances in Battery Technologies for Electric Vehicles*, ed. by B. Scrosati, J. Garcia, W. Tillmetz, (Woodhead Publishing Series in Energy, Woodhead), pp. 173–190
10. RTA Transit, Howard County Electric Bus Project: Interactive Electric Bus. RTA Transit. August 23, 2018. http://www.transitrta.com/interactive/
11. California HVIP, Hybrid and Zero-Emission Truck and Bus Voucher Incentive Project (HVIP). California HVIP. September 3, 2020. https://www.californiahvip.org/
12. Proterra, Catalyst: 40 Foot Bus Performance Specifications. Proterra. https://www.proterra.com/wp-content/uploads/2016/08/Proterra-Catalyst-Vehicle-Specs.pdf, 2018
13. A.E. Trippe, R. Arunachala, T. Massier, A. Jossen, T. Hamacher, Charging optimization of battery electric vehicles including cycle battery aging, in *IEEE PES Innovative Smart Grid Technologies, Europe*, 1–6 (2014). https://doi.org/10.1109/ISGTEurope.2014.7028735
14. B. Lunz, Z. Yan, J.B. Gerschler, D.U. Sauer, Influence of Plug-in Hybrid Electric Vehicle Charging Strategies on Charging and Battery Degradation Costs. Energy Policy **46**(July), 511–519 (2012). https://doi.org/10.1016/j.enpol.2012.04.017
15. J. Hanlin, Battery electric buses smart deployment. Presented at the Zero Emission Bus Conference, November 30, 2016. http://www.cte.tv/wp-content/uploads/2016/12/5_Hanlin.pdf
16. K. Tomford, Chicago Transit Authority (CTA) Electrification of Mass Transit. Presented at the NextGrid Illinois, February 14, 2018
17. L. Eudy, M. Jeffers, *Foothill Transit Battery Electric Bus Demonstration Results: Second Report. NREL/TP-5400-67698* (NREL, Golden, 2017)
18. L. Engel, AVTA's Journey to Fleet Electrification. Presented at the BusCon, Indianapolis, September 11, 2017. https://eventsimages.bobitstudios.com/upload/pdfs/bcs/2017/speaker-presentations/alook-at-conductive-inductive-charging-of-electric-buses-len-engel-91217-945am.pdf
19. W.F. Torrey, D. Murray, *An Analysis of the Operational Costs of Trucking: 2015 Update.* Arlington, Virginia (2015a)
20. U.S. Environmental Protection Agency, Draft Inventory of U.S. Greenhouse Gas Emissions and Sinks: 1990–2014 (2016)
21. W. Feng, M.A. Figliozzi, Conventional vs electric commercial vehicle fleets: a case study of economic and technological factors affecting the competitiveness of electric commercial vehicles in the USA. Procedia Soc. Behav. Sci. **39**, 702e711 (2012). https://doi.org/10.1016/j.sbspro.2012.03.141

22. MJB and A, Comparison of Modern CNG, Diesel and Diesel Hybrid-electric Transit Buses. Concord, MA (2013)
23. F. Tong, P. Jaramillo, M.L. Azevedo, Comparison of life cycle greenhouse gases from natural gas pathways for medium and heavy-duty vehicles. Environ. Sci. Technol. **49**, 7123e7133 (2015). https://doi.org/10.1021/es5052759
24. Transportation Sector Energy Use by Mode and Type (US Energy Information Administration, 2020); https://www.eia.gov/outlooks/aeo/data/browser/#/?id=45-AEO2020&sourcekey=0
25. N. Williams, D. Murray, *An Analysis of the Operational Costs of Trucking: 2020 Update* (American Transportation Research Institute, 2020)
26. J. Thomas, Drive cycle powertrain efficiencies and trends derived from EPA vehicle dynamometer results. SAE Int. J. Passeng. Cars Mech. Syst. **7**, 1374–1384 (2014)
27. E. Çabukoglu, G. Georges, L. Küng, G. Pareschi, K. Boulouchos, Battery electric propulsion: An option for heavy-duty vehicles? Results from a Swiss case-study. *Transport. Res. C* **88**, 107–123 (2018)
28. *Freight Transportation Energy Use* (US Energy Information Administration, 2020). Accessed 11 Jan 2022. https://www.eia.gov/outlooks/aeo/data/browser/#/?id=58-AEO2020&cases=ref2020&sourcekey=0
29. H. Zhang, W. Chen, W. Huang, TIMES modelling of transport sector in China and USA: comparisons from a decarbonization perspective. Appl. Energy **162**, 1505–1514 (2016)
30. M. Muratori et al., Role of the freight sector in future climate change mitigation scenarios. *Environ. Sci. Technol.* **51**, 3526–3533 (2017)
31. L.H. Kaack, P. Vaishnav, M.G. Morgan, I.L. Azevedo, S. Rai, Decarbonizing intraregional freight systems with a focus on modal shift. *Environ. Res. Lett.* **13**, 083001 (2018)
32. S. Yeh et al., Detailed assessment of global transport-energy models' structures and projections. *Transport. Res. D* **55**, 294–309 (2016)
33. S.C. Davis, R.G. Boundy, Transportation Energy Data Book Edition 38.1 (Oak Ridge National Laboratory, 2020.); https://tedb.ornl.gov
34. Ford, Home Depot Talk Electrifying Fleets, Cutting Transport Emissions, Raising Revenue. Accessed 11 Jan 2022. https://www.environmentalleader.com/2017/01/ford-home-depot-talk-electrifying-fleets-cutting-transport-emissions-raising-revenue/
35. J.R. Pierson, R.T. Johnson, the battery Designer's challenge — satisfying the ever-increasing demands of vehicle electrical systems. J. Power Sources **33**, 309–318 (1991). https://doi.org/10.1016/0378-7753(91)85069-9
36. Peter Weidenhammer "A sports car that covers over 300 miles with superb performance—but without a drop of gasoline? Welcome to the future: the Mission E electric concept car"- Porshe cars North America 2019 [Online]. Available: https://www.porsche.com/usa/aboutporsche/christophorusmagazine/archive/374/articleoverview/article01/. Accessed 2 Feb 2022
37. J. Winterhagen, "Mission E"- Dr. Ing. h.c. F. Porsche AG 2018[Online]. Available: https://newsroom.porsche.com/en/products/porsche-taycan-mission-e-drive-unit-battery-charging-electro-mobilitydossier-portscar-production-christophorus-387-15827.html. Accessed 2 Feb 2022
38. M. Alatalo "Module size investigation on fast chargers for BEV" Swedish Electromobility Centre, 2018 [Online]. Available: http://emobilitycentre.se/wp-content/uploads/2018/08/Module-sizeinvestigation-on-fast-chargers-for-BEV-final.pdf. Accessed 22 Dec 2018
39. Tritium, VEEFIL-RT 50KW DC FAST CHARGER, Tritium Pty Ltd 2019 [Online]. Available: https://www.tritium.com.au/product/productitem?url=veefil-rt-50kwdc-fast-charger. Accessed 2 Feb 2022
40. Tritium, LAUNCHING SOON- VEEFIL-PK DC ULTRA-FAST CHARGER 175 kW- 475 kW, Tritium Pty Ltd 2019 [Online]. Available: https://www.tritium.com.au/veefillpk. Accessed 2 Feb 2022
41. EVBOX, LAUNCHING SOON- VEEFIL-PK DC ULTRA-FAST CHARGER 175 kW- 475 kW, EvBox 2019 [Online]. Available: https://www.evbox.fr/produits/borne-recharge-rapide. Accessed 2 Feb 2022

42. ABB, High Power Charging, ABB 2019 [Online]. Available: https://new.abb.com/ev-charging/products/car-charging/high-powercharging. Accessed 2 Feb 2022
43. ChargePoint, ChargePoint Express Plus, ChargePoint 2019 [Online]. Available.: https://www.chargepoint.com/products/commercial/express-plus. Accessed 22 Feb 2022
44. EVteQ, SET-QM EV Charging Module, Evteq 2017 [Online]. Available http://www.evteq-global.com/product-details/set-qm-evcharging-module. Accessed 2 Feb 2022
45. Siemens, High power charging for current and future ecars, Siemens Mobility2018 [Online]. Available: https://www.siemens.com/global/en/home/products/mobility/road-solutions/electromobility/ecars-high-power-charginginfrastructure.html. Accessed 2 Feb 2022
46. Siemens, Sinamics DCP, Siemens AG 2019 [Online]. Available: https://w3.siemens.com/drives/global/en/converter/dc-drives/dc-cconverters/pages/sinamics-dcp.aspx. Accessed 2 Feb 2022
47. PRE, 12.5 kW Charger Module, PRE 2018 [Online]. Available: http://www.prelectronics.nl/media/documenten//12k5W_Charger_datasheet_V11.180417.194409.pdf. Accessed 2 Feb 2022
48. PRE, 25 kW DC/DC Charger Module, PRE, 2018[Online]. Available: http://www.prelectronics.nl/media/documenten//25kW_Charger_datasheet_V13.180417.194545.pdf. Accessed 2 Feb 2022
49. Phihong, Phihong EV Chargers, Phihong 2017 [Online]. Available: https://www.phihong.com.tw/newcatalog/2017%20EV%20Chargers%20%E5%86%8A%E5%AD%90_20170420_EN.pdf. Accessed 2 Feb 2022
50. Ies, Charger Module, IES Synergy [Online]. Available: http://www.ies-synergy.com/en/products/keywatt-charging-stationspower-modules/charger-module. Accessed 2 Feb 2022
51. D. Cericola, R. Kötz, Hybridization of rechargeable batteries and electrochemical capacitors: principles and limits. Electrochim. Acta **72**, 1–17 (2012). https://doi.org/10.1016/j.electacta.2012.03.151
52. S. Manzetti, F. Mariasiu, Electric vehicle battery technologies: from present state to future systems. Renew. Sustain. Energy Rev. **51**, 1004–1012 (2015). https://doi.org/10.1016/j.rser.2015.07.010
53. M. Hannan, F. Azidin, A. Mohamed, Hybrid electric vehicles and their challenges: A review. Renew. Sustain. Energy Rev **29**, 135–150 (2014). https://doi.org/10.1016/j.rser.2013.08.097
54. M. Hannan, M. Hoque, A. Mohamed, A. Ayob, Review of energy storage systems for electric vehicle applications: issues and challenges. Renew. Sustain. Energy Rev. **69**, 771–789 (2017). https://doi.org/10.1016/j.rser.2016.11.171
55. A. Eftekhari, Low voltage anode materials for lithium-ion batteries. Energy Storage Mater **7**, 157–180 (2017). https://doi.org/10.1016/j.ensm.2017.01.009
56. A. Eftekhari, LiFePO4/C nanocomposites for lithium-ion batteries. J. Power Sources **343**, 395–411 (2017). https://doi.org/10.1016/j.jpowsour.2017.01.080
57. H. Shi, Activated carbons and double layer capacitance. Electrochim. Acta **41**, 1633–1639 (1996). https://doi.org/10.1016/0013-4686(95)00416-5
58. R. Kötz, M. Carlen, Principles and applications of electrochemical capacitors. Electrochim. Acta **45**, 2483–2498 (2000). https://doi.org/10.1016/S0013-4686(00)00354-6
59. A. Burke, Ultracapacitors: why, how, and where is the technology. J. Power Sources **91**, 37–50 (2000). https://doi.org/10.1016/S0378-7753(00)00485-7
60. A. Pandolfo, A. Hollenkamp, Carbon properties and their role in supercapacitors. Power Sources **157**, 11–27 (2006). https://doi.org/10.1016/j.jpowsour.2006.02.065
61. J.P. Sharma, T. Bhatti, A review on electrochemical double-layer capacitors. Energy Convers. Manage **51**, 2901–2912 (2010). https://doi.org/10.1016/j.enconman.2010.06.031
62. A. González, E. Goikolea, J.A. Barrena, R. Mysyk, Review on supercapacitors: technologies and materials. Renew. Sustain. Energy Rev. **58**, 1189–1206 (2016). https://doi.org/10.1016/j.rser.2015.12.249

Chapter 5
Fast-Charging Infrastructure for Transit Buses

Hossam A. Gabbar and Mohamed Lotfi

Nomenclature

BS	Battery swap
CC	City Center
DC	Depot charger
DER	Distributed energy resource
EB	Electric bus
ESS	Energy storage system
EV	Electric vehicle
FC	Flash charger
HF	High frequency
LD	Long distance
LF	Low frequency
LV	Low voltage
SD	Short distance
SoC	State-of-charge
SU	Suburban
TC	Terminal charger

The original version of this chapter was revised. The correction to this chapter is available at https://doi.org/10.1007/978-3-031-09500-9_18

H. A. Gabbar (✉)
Faculty of Energy Systems and Nuclear Science, Ontario Tech University (UOIT), Oshawa, ON, Canada

Faculty of Engineering and Applied Science, Ontario Tech University, Oshawa, ON, Canada
e-mail: Hossam.gabbar@uoit.ca

M. Lotfi
Faculty of Energy Systems and Nuclear Science, Ontario Tech University (UOIT), Oshawa, ON, Canada

© Springer Nature Switzerland AG 2022, Corrected Publication 2023
H. A. Gabbar, *Fast Charging and Resilient Transportation Infrastructures in Smart Cities*, https://doi.org/10.1007/978-3-031-09500-9_5

Sets and Indexes

i Index for stops
j Index for trips
k Index for buses
r Index for routes

Index of the First Element in a Set

end Index of the last element in a set
I^r Set of all stops in route r
J^r Set of all trips in route r
K^r Set of all buses in route r
R Set of all routes
H Set of all available on-route charger types

Variables

d_r^s Average distance between stops on route r
L_r Length of route r
N_r^s Number of stops in route r
d_r^d Average daily distance between stops on route r
H_r Operating hours of route r (time difference between first and last buses of the day)
T_r Average duration of route r
N_r^t Daily number of trips for route r
bs_r Binary variable indicating battery swap depot existence in route r
bd_r Binary variable indicating bus depot existence in route r
$b_{k,r}$ Capacity of battery on bus k on route r
$x_{i,h,r}$ Binary variable indicating presence of on-route charger type h, at stop i, in route r
$e_{i,j,r}$ Energy charged at stop i, during trip j, in route r
d_{year} Number of days in a year
n_r^{bus} Number of buses deployed to route r
F_r^b Frequency of buses is route r
$E_{i,j,k,r}$ Battery SoC at stop i during trip j, on bus k, in route r
$\overline{B}$ Upper bound for the batteries' SoC
$\underline{B}$ Lower bound for the batteries' SoC
$\overline{E}_h$ Maximum charge capacity of on-route charger type h
$\overline{E}_{DC}$ Maximum charge capacity of the depot charger

5.1 Introduction

Electrification of buses is widely recommended to reduce greenhouse gas (GHG) emissions from conventional fossil fuel buses. There are a number of challenges to achieve high-performance charging infrastructures for electric buses (EBs). Time to charge EB during the daily trips should be minimum to reduce waiting time and passengers' trip time. There are a number of challenges that face the transition of electric bus fleet, including planning charging stations, adopting fast- and ultrafast-charging stations, and possible business models of swapping battery in EBs to avoid waiting time to charge on-route [1]. The deployment of electric buses in different regions, such as Europe, reflected challenges including infrastructure planning, marketplace, pricing, charging infrastructure, and business models [2]. The performance of charging infrastructure is evaluated with a number of performance measures, such as cost, time, mobility, and social factors. The optimization of charging infrastructures is essential to achieve profitable transportation electrification [3]. The charging stations are interfaced with the grid, where charging demands affect the grid performance. Hence, the study of charging station interface to the grid is critical to meet charging demand profiles and grid performance [4]. The expansion of transportation electrification and charging infrastructures requires proper analysis of grid impacts to balance charging load profiles with grid condition [5]. The deployment strategies and planning of fast-charging stations should consider electrification load profiles [6]. The coupling between transportation electrification and grid condition will support the planning of large-scale charging stations [7]. The charging performance could be enhanced with different strategies such as prediction of charge ahead of time [8]. The overall performance of electric buses could be enhanced with smart charging capabilities where coordination between buses and stations, among buses, and stations, and the grid could provide enhanced performance [9]. There are a number of bus charging technologies which are available in the market, such as the technologies from ABB [10]. The different deployment strategies opened the door for implementation projects in different regions, city centers, urban areas, suburban areas, and remote communities. To achieve successful installation projects of charging infrastructures, the requirements should be analyzed in terms of energy storage, mobility demand, and social factors [11]. And the bus route planning should also be considered as an integral part of the charging infrastructure. Case study is analyzed for bus electrification in Porto [12] and in London [13]. To reduce the impacts on the grid, renewable energy sources, such as solar and wind, are widely used and integrated with charging stations in large-scale stations and small-scale stations in parking lots [14]. The scheduling of bus routes should be planned properly to optimize the overall transit performance whether by adopting central charging strategies with battery swapping or by charging on-routes [15]. Computational intelligence techniques could be utilized to enhance energy management of electric bus charging performance with deep learning techniques where selection of nearest station based on battery state could be optimized [16]. Stochastic learning techniques are also utilized to improve energy management of electric bus charging and the overall performance [17]. In order to have better understanding of different techniques and strategies for the planning of electric bus deployment and charging infrastructure, it is important to provide analysis using bus transit and charging infrastructure models. This chapter will present possible models and associated parameters for electric bus charging with different strategies and scenarios.

5.2   Electric Bus Charging Models

There are a number of EB charging styles that can be selected based on different trips, regions, technologies, and techno-economic preferences. Figure 5.1 shows a public transit network with possible charging models for EB, which include on-route charging using flash charger, bus terminal charging, or overnight charging in bus depot. Also battery swapping could be implemented in selected charging stations.

The buses are parked in the depots when they are not in service. Buses usually stop for longer periods and can have longer charging time. Fast or ultrafast charging will be implemented in charging station on-route via flash charging.

The depot charger (DC) is used to charge buses when they are out of service and parked at the depot, which is usually off-route. The power rate of DC is typically in the power range from 50 to 100 kW, which are usually used for slow charging, that is, overnight or when they are out of service. The terminal charger (TC) is typically installed for on-route charging at main terminals. Buses stay a few minutes at the TC station. The rated power of TC is ranging from 500 to 600 kW. TC is commonly connected to high-voltage power grid. The flash charger (FC) is used for on-route fast charging at regular bus stops. FC has a rated power ranging from 400 to 500 kW,

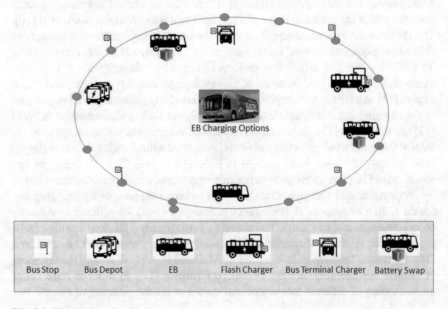

Fig. 5.1   Electric bus charging on-route

which is connected to low-voltage (LV) power grid. Batteries are commonly used with FC to reduce load spikes on the LV grid. Bus stop at regular stops for a few seconds; hence, fast charging is required.

d_r^d is the average daily distance on route r, which is calculated based on total operating hours per day, trip average time, and trip average distance. It is defined using Eq. 5.1:

$$d_r^d = \frac{H_r}{T_r} \cdot L_r = N_r^t \cdot L_r \qquad (5.1)$$

d_r^s is the average distance between stops for route r, which is calculated based on the total distance on route r, defined as L_r, and number of stops on route r, based on Eq. 5.2:

$$d_r^s = \frac{L_r}{N_r^s - 1} \qquad (5.2)$$

The proposed EB charging infrastructure planning mechanism is based on defining number of scenarios where different combinations of charger types on a given route can serve number of buses on the same route. The selection of the best charger type is based on optimization model in view of performance measures, which are defined in the following section.

5.3   Performance Measures

The performance of the bus transit network is evaluated based on multiple performance measures, as described in Table 5.1.

KPI-C	Cost	KPI-C1: Capital cost of charging infrastructure
		KPI-C2: Cost of daily trips
		KPI-C3: Cost of daily energy back to grid (V2G)
KPI-T	Time	KPI-T1: Average time of daily trips
		KPI-T2: Average charging time of daily trips
		KPI-T3: Total delay time for daily trips
KPI-B	Battery	KPI-B1: Battery lifetime
KPI-I	Charging infrastructure	KPI-I1: Energy not served for charging buses
		KPI-I2: Total daily energy back to the grid (V2G)
		KPI-I3: Reliability of charging infrastructure
KPI-S	Social	KPI-S1: Mobility density per day
		KPI-S2: Area coverage index
KPI-R	Risk	KPI-R1: Risk value of not reaching charging station in normal condition
		KPI-R2: Risk value of not reaching charging station in abnormal condition

Table 5.1 Specifications of routes for the case study

	Route A	Route B	Route C
Number of trips (per day)	7	14	16
Number of stops (per trip)	70	80	75
Total number of stops (per day)	350	800	1200
Trip length (km)	25	20	15
Bus size (m)	18	24	24
Average consumption (kWh/km)	1.8	2.2	2.2

The proposed performance measures include cost measures related to capital and operating costs associated with charging infrastructures. Time factors are considered as part of the overall performance of charging infrastructures, including charging time, trip time, and delay time in each trip. The performance of charging infrastructure is also monitored and optimized in terms of energy not served to charge incoming buses, total energy back to the grid (in different peak times), and the reliability of the charging infrastructure. Social factors are considered in terms of mobility density per day and area coverage index to ensure equity for reduced delays in different regions. The risk factors are also monitored in terms of the risk of the bus not being able to reach the next charging station with empty battery in normal and abnormal conditions.

5.4   Case Study

To understand the proposed modeling of charging infrastructures for transit buses, case studies are illustrated in this section. Table 5.1 shows specifications for the case study represented by three different routes. Route A has 7 trips per day, route B has 14 trips per day, and route C has 16 trips per day. The case study shows different parameters for the three routes in terms of stops per trip, stops per day, trip length, bus size, and average energy consumption per Km.

The understanding of the routes is used to analyze the techno-economic specifications of the chargers in each route. Table 5.2 shows different charger classifications, models, rated power, maximum charging time, capital cost of the charger, operating cost of each charger, and lifetime of the charger. These parameters are used to analyze and optimize charging infrastructure in terms of defining charger model, type, size, and location with respect to bus stops. The different scenarios will be optimized in view of key performance indicators defined in Table 5.1.

The selection of batteries for the electric buses will influence the selection of charging infrastructure specifications. Table 5.3 shows different techno-economic parameters defined for batteries of electric buses, including capital cost of the battery, operating cost, lifetime, and state-of-charge upper and lower limits. These battery parameters will be used to analyze and optimize battery selection within charging infrastructures.

Table 5.2 Techno-economic specifications of chargers

	DC	TC		FC	
Charger classification	Depot	On-route		On-route	
Model	Standard	Slow	Fast	Slow	Fast
Rated power (kW)	50	400	600	400	600
Maximum charging time	5 h	3 min		10 s	
Capital cost (EUR)	100 k	290 k	310 k	320 k	320 k
Operating cost (EUR/year)	120	2100		2100	
Lifetime (years)	20	20		20	

Table 5.3 Techno-economic specifications of batteries

Capital cost (EUR/kWh)	250
Operating cost (EUR/year)	–
Battery lifetime (years)	5
State-of-charge upper boundary (%)	90
State-of-charge lower boundary (%)	10

The proposed charging infrastructures for transit buses are useful and comprehensive to enable detailed analysis of different deployment strategies and operational scenarios based on user requirements and target performance. Optimization methods could be used to maximize profits and the overall performance of charging infrastructures for transit buses.

5.5 Summary

This chapter presented detailed models for charging infrastructure to support electrification of transit buses. Different routes are defined in terms of trips, bus technologies, charging technologies, battery technologies, and performance measures. Case study specifications for routes, chargers, and battery technologies are defined as basis for the analysis of charging infrastructures for transit buses.

Acknowledgments Authors would like to thank members in the Smart Energy Systems Laboratory (SESL).

References

1. S. Pelletier, O. Jabali, J.E. Mendoza, G. Laporte, The electric bus fleet transition problem. Transp. Res. Part C Emerg. Technol. **109**(October), 174–193 (2019)
2. L. Mathieu, *Electric Buses Arrive on Time – Marketplace, Economic, Technology, Environmental and Policy Perspectives for Fully Electric Buses in the EU* (Transport & Environment, 2018)

3. M. Rogge, E. Van Der Hurk, A. Larsen, D.U. Sauer, Electric bus fleet size and mix problem with optimization of charging infrastructure. Appl. Energy **211**, 282–295 (2018)
4. Z. Wu, F. Guo, J. Polak, G. Strbac, Evaluating grid-interactive electric bus operation and demand response with load management tariff. Appl. Energy **255**(August), 113798 (2019)
5. M. Bhaskar Naik, P. Kumar, S. Majhi, Smart public transportation network expansion and its interaction with the grid. Int. J. Electr. Power Energy Syst. **105**(December 2017), 365–380 (2019)
6. Y. He, Z. Song, Z. Liu, Fast-charging station deployment for battery electric bus systems considering electricity demand charges. Sustain. Cities Soc. **48**(October 2018), 2019
7. Y. Lin, K. Zhang, Z.-J.M. Shen, B. Ye, L. Miao, Multistage large-scale charging station planning for electric buses considering transportation network and power grid. Transp. Res. Part C Emerg. Technol. **107**(August), 423–443 (2019)
8. The European electric bus market is charging ahead, but how will it develop? | McKinsey. [Online]. Available https://www.mckinsey.com/industries/oil-and-gas/our-insights/the-european-electric-bus-market-is-charging-ahead-but-how-will-it-develop. Accessed 05 Mar-2020
9. IRENA, *Innovation Outlook: Smart Charging for Electric Vehicles* (IRENA, 2019)
10. ABB Canada, *Electrification of Public Transport*, Toronto (2018)
11. M. Rogge, S. Wollny, D. Sauer, Fast charging battery buses for the electrification of urban public transport – A feasibility study focusing on charging infrastructure and energy storage requirements. Energies **8**(5), 4587–4606 (May 2015)
12. D. Perrotta et al., Route planning for electric buses: A case study in oporto. Procedia Soc. Behav. Sci. **111**, 1004–1014 (Feb. 2014)
13. Transport for London, *Key Bus Routes in Central London* (2020). [Online]. Available http://content.tfl.gov.uk/bus-route-maps/key-bus-routes-in-central-london.pdf. Accessed 06 Mar 2020
14. P. Nunes, R. Figueiredo, M.C. Brito, The use of parking lots to solar-charge electric vehicles, in *Renewable and Sustainable Energy Reviews*, vol. 66, (Elsevier Ltd, 2016), pp. 679–693
15. Q. Kang, J. Wang, M. Zhou, A.C. Ammari, Centralized charging strategy and scheduling algorithm for electric vehicles under a battery swapping scenario. IEEE Trans. Intell. Transp. Syst. **17**(3), 659–669 (Mar 2016)
16. H. Tan, H. Zhang, J. Peng, Z. Jiang, Y. Wu, Energy management of hybrid electric bus based on deep reinforcement learning in continuous state and action space. Energy Convers. Manag. **195**(January), 548–560 (2019)
17. Z. Chen, L. Li, X. Hu, B. Yan, C. Yang, Temporal-difference learning-based stochastic energy management for plug-in hybrid electric buses. IEEE Trans. Intell. Transp. Syst. **20**(6), 2378–2388 (2019)

Chapter 6
A Robust Decoupled Microgrid Charging Scheme Using a DC Green Plug-Switched Filter Compensator

Mustafa Ergin Şahin and Adel Mahmoud Sharaf

6.1 Introduction

The renewable photovoltaic sources and their emerging use in energy storage applications for battery and hybrid storage devices are essential with time and research by more scientists [1–3]. The power electronic converters and their controllers are the fundamental parts of the hybrid storage systems. These components transmit the generated energy from PV sources to storage devices or the grid [4, 5]. Also, some FACTS (flexible AC transmission system) components increase the efficiency of the general system. For example, a DC green plug FACTS which includes switched capacitor compensated scheme is proposed with a converter to increase the general system efficiency in some applications [6, 7].

PV array operating efficiency changes based on solar irradiation and temperature level. Also, these sources are nonlinear and act depending on the current-voltage (*I-V*) characteristic of the photovoltaic source. This nonlinearity is affected by the efficiency of the PV panel [8]. Required to work at the maximum power point (MPP) of photovoltaic panels to increase the efficiency, and it is possible using some controllers and converters to connect the PV panel's output [5, 9]. Also, it is required to supply joint bus stabilization for efficient battery charging. A supercapacitor (SC) and FACTS can be proposed as a solution [9–11]. A novel FACTS-based distributed

The original version of this chapter was revised. The correction to this chapter is available at https://doi.org/10.1007/978-3-031-09500-9_18

M. E. Şahin (✉)
Department of Electrical and Electronics Engineering, Recep Tayyip Erdoğan University, Rize, Turkey

A. M. Sharaf
Sharaf Energy Systems, Inc- Fredericton, New Brunswick, Canada

green plug-switched filter compensator (GPSFC) system is given for microgrid-connected wind energy systems in a paper [12]. In another study, a FACTS-based dynamic switched capacitor-type filter (DSCTF) compensator scheme is presented with different control methods and loads for distribution systems [13]. To improve the damping of oscillations in power systems, supplementary control laws are applied to the developed FACTS devices which are known as power oscillation damping (POD) control [14].

The research presents many error-driven control methods for a battery-powered combined hybrid system for electric vehicles (EVs). The proposed regulation schemes include modified PID control strategies and a sliding mode control (SMC) scheme for dynamic variable structure [15]. Conversely, intelligent control system developments and their microgrid applications are discussed in a book chapter [16]. Moreover, a modified hierarchical multistage fuzzy logic and PID control scheme for a hybrid supplied AC grid driven by PV and battery charging system is presented in another paper [17]. A practical single-phase induction motor controller based on the multi-objective genetic algorithm with some green plug-switched filter compensator schemes is given with low-cost compensator schemes [18]. A new flexible and precisely controlled self-adjusting PV and battery-powered light-emitting diode (LED) lighting system using sinusoidal pulse width modulation (SPWM) switching are presented in another paper. A dual-loop error-driven control in this scheme to supply the LED lighting load is a weighted and modified PID control scheme for the PV and battery source [19, 20].

The paper presents a battery charging scheme using a DC green plug filter PWM-switched multi-loop hybrid regulation controller. This controller uses the battery, PV, and common bus DC voltages and currents to control loop error signals and drive PID controllers. The PID outputs are used to generate PWM switching signals. The main idea and proposed PV-powered green plug schemes are given in this paper, as shown in Fig. 6.1. The controller design steps and controller components are given more detail in this study. The digital simulation models are designed using MATLAB/Simulink software, and the digital simulation results are presented for some operating conditions of two schemes. Also, the harmonic analyses for this scheme are investigated in more detail.

After the introduction part, the proposed efficient PV-powered schemes are given in Sect. 6.2. The controller design steps and structure are given in Sect. 6.3. The digital simulation results are presented in Sect. 6.4, and the conclusions are given in Sect. 6.5.

6.2 The Proposed Efficient PV-Powered Schemes

The paper presents a PV-powered scheme with two proposed efficient low ripple two green plug schemes for battery charging using green plug-supercapacitor DC interface filter schemes, as shown in Fig. 6.1a, b. The green plugs act as super storage and DC side filter to reduce voltage transients and DC inrush current conditions. The operation uses MOSFET/IGBT switched green plug DC capacitors controlled by the

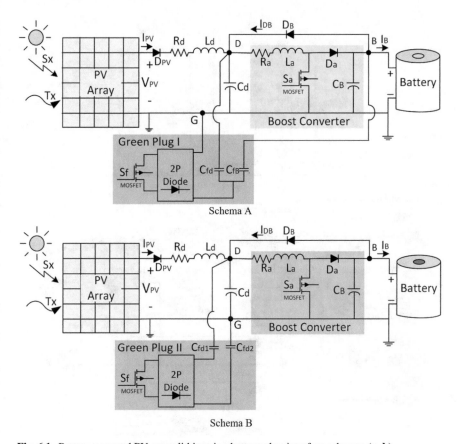

Fig. 6.1 Battery-powered PV-green lithium-ion battery plug-interface schemes (**a, b**)

dual regulating decoupled and descaled controller. The freewheeling energy recovery diode D_B ensures energy exchanges between PV source and lithium-ion battery loads during energization/de-energization. The boost type DC-DC converter is controlled using an integrated volt-current-power (V-I-P) weighted charging controller with assigned weighting. It ensures descaling and best hybrid fast-charging modes using weighted, decoupled, and descaled controllers to reduce voltage transients and inrush current conditions on the DC side and ensure energy-efficient operation.

The proposed green plug-switched schemes are combined with a PV model, a boost converter, battery, and green plug-switched filter compensator (GPSFC) components. Derived from the basic Shockley diode equation, the PV cell basic equations for the equivalent circuit and the PV model are designed as shown in Fig. 6.2a. The PV current for the series- and parallel-connected PV cells is given in Eq. 6.1, and voltage is given in Eq. 6.2. Using these two equations, the PV source model is designed for desired arrays in MATLAB/Simulink software, as shown in Fig. 6.2b. This PV model is designed, tested, verified, and used previously by the authors [19,

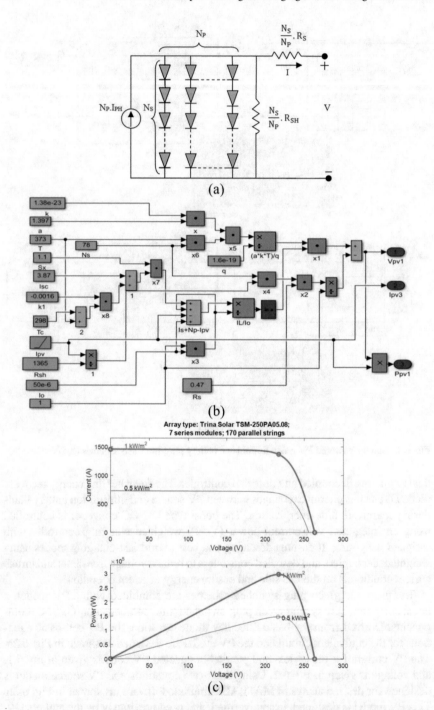

Fig. 6.2 The basic models of the designed PV scheme, (**a**) PV equivalent circuit model, (**b**) PV source model in MATLAB/Simulink, and (**c**) obtained *I-V* and *P-V* characteristics of the PV model

20, 23, 24]. For the developed model, the current-voltage (I-V) and the power-voltage (P-V) characteristics are given for 220 V voltage and 300 kW maximum power during the 1 kW/m³ (S_x) solar irradiation level as shown in Fig. 6.2c:

$$I = N_P.I_{PH} - N_P.I_0 \left[\exp\left(q \left(\frac{V}{N_S} + \frac{I.R_S}{N_P} \right) / k.T_C.A \right) - 1 \right] - \left(\frac{N_P.V}{N_S} + I.R_S \right) / R_{SH} \quad (6.1)$$

$$V_{PV} = \frac{N_S.n.k.T}{q} \ln \left[\frac{\left(I_{SC} + K_I \left(T - T_{ref} \right) \right).G + N_P I_0 - I_{PV}}{I_0.N_P} \right] - \frac{N_S}{N_P}.\frac{R_s}{R_{sh}}.I_{PV} \quad (6.2)$$

The GPSFC includes switchable capacitors, which are introduced at the DC bus to improve voltage stabilization of the DC bus and the efficiency of the DC microgrid. To regulate the DC bus voltage, GPSFC is used. Also, it is used to minimize inrush current transients for PV and battery nonlinear I-V characteristics. The controller compares the voltage of the DC bus value with a reference voltage value and generates the switching pulses for filter switches (S_f) by the error value [11]. The proposed green plug models are given in Fig. 6.3a, b. The first GPSFC is connected between the boost converters' input and output terminals to compensate for the boost converters' filter capacitors (C_d, C_B). The second GPFCS is connected only between the boost converters' input positive poles to the ground to compensate for the boost converters' input filter capacitor.

The central part of this scheme is the boost converter required to increase and regulate the PV source voltage to energize the battery or other loads. This switching converter includes a series inductance with the source and increases the input voltage at the output with the other components. The DC-to-DC boost converter circuit with the current flow and the inductance voltage and current variation depending on different duty cycles (D) of switching signal are given in Fig. 6.4a, b. This converter working principle depends on the switching component (S) states. The input current flows through the inductance when the switch is open (S_{ON}). The inductance current flows through the diode, capacitor, and load components to complete the circle with a decreasing inductance current when the switch is closed (S_{OFF}), as seen in Fig. 6.4a.

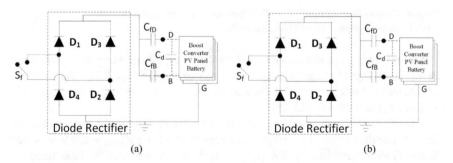

(a)   (b)

Fig. 6.3 Green plug-switched filter compensator model I (a) and model II (b)

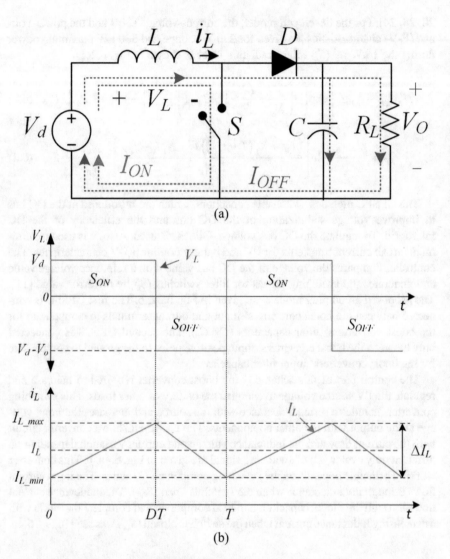

Fig. 6.4 Current flow states of DC-to-DC boost converter circuit and (**a**) inductance voltage, and current variations depend on switching states (**b**)

The output voltage of the converter boosts the input voltage every time. When the switch is *on*, input voltage (V_d) equals inductance voltage (V_L). When the switch is *off*, the output voltage is equal to the sum of input (V_d) and inductance voltage (V_L), as given in Fig. 6.4b [21, 22, 25].

The minimum value of the converter inductor can be formulated and found using Eq. 6.3. The duty ratio ($D = 1 - V_d/V_o$) is accepted $D = 0.5 \times V_{d(min)} = 220$ V for small S_X and T_X values and battery SOC. $f_S = 2500$ Hz and $\Delta I_L = 68.2$ A for limit value:

$$\Delta I_L = \frac{V_{d(\min)} \times D}{f_s \times L} = \frac{220 \times 0.5}{2500 \times L} = 68.2 \rightarrow L = \frac{110}{170500} = 645\,\mu H \text{ and } L > 645\,\mu H$$

$$(6.3)$$

The minimum value of the output capacitor of the converter for the under-required ripple on the output voltage can be formulated and defined using Eq. 6.4. The maximum ripple value of the capacitor voltage accepted is less than 1% for the output voltage. These limit values can be increased according to circuit whole conditions:

$$C_{\min} = \frac{I_{out(\max)} \times D}{f_s \times \Delta V_o} = \frac{400 \times 0.5}{2500 \times 2.2} = \frac{200}{5500} = 36363\,\mu F \text{ and } C > 36363\,\mu F \qquad (6.4)$$

The battery model is designed using MATLAB/Simulink software, as given in Fig. 6.5a. The battery model uses the lead-acid, lithium-ion, and other types of battery equations and adjusts the nominal voltage, the rated capacity, and the initial state-of-charge (SoC) response time as shown in Fig. 6.5b. Nominal current discharge characteristics for the battery model parameters are given in Fig. 6.5c as ampere hour for nominal voltage [27–30].

6.3   The Controller Design Steps and Structure

The dual regulating controller shown in Fig. 6.6 is a coupled version using energy freewheeling exchanges from the load battery side and the PV input side with the green plug scheme to reduce transients and inrush conditions on the DC system. It is essential to ensure all loops' best weightings reduce interactions that may result in sudden inrush DC transients by describing and enhancing major and minor loop functions in the battery charging regulator controlling. The PWM switched DC-DC boost converter for lithium-ion battery charging load and the second regulator controlling the DC green plug filter PWM. The weighted extra modulation loop for the intercoupled freewheeling exchange current error (eI_{DB}) is added to ensure a proper energy balance of DC power exchanges.

The *dual loop_1* and *dual loop_2* regulators, current boost supplementary loop regulators, and PID controllers are simulated in Simulink software as seen in Fig. 6.7a–d. This loop generates error signals (e_1, e_{t2}, e_{ldb}) that drive PID_A and PID_B controllers. The current loop regulators' error signal is added with dual-loop regulators' error signals to amplify the error signals. At last, the two PID controller output signals are used to generate switching PWM 1 and switching PWM 2 signals. One of these switching signals (S_C) drives the converter; the other (S_f) drives the green plug filter compensator.

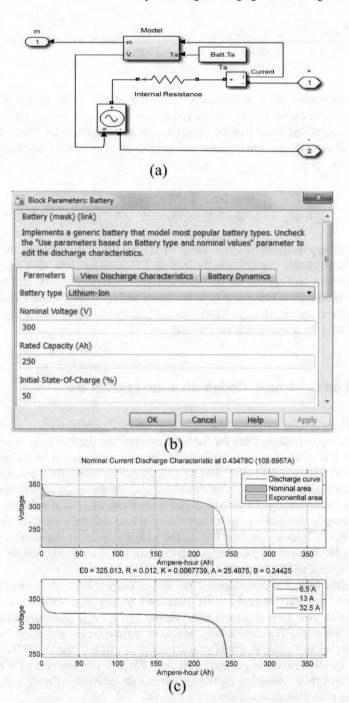

Fig. 6.5 (a) The battery model, (b) battery interface in MATLAB/Simulink, and (c) current discharge characteristics for the battery model parameters

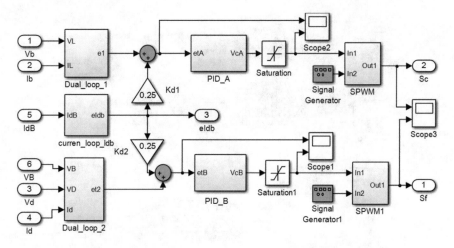

Fig. 6.6 The control structure with the dual regulation loop-decoupled controllers

The error signals for dual loops 1 and 2 are designed using Eqs. 6.5 and 6.6. The current error boosting supplementary circle is designed using Eq. 6.7. These error signals are applied to the PID_A and PID_B controllers using Eqs. 6.8 and 6.9:

$$e_1(t) = \left(V_{L\_ref} - \frac{V_L(t)}{V_{d\_base}} \left(\frac{1}{sT_1 + 1} \right) \right) K_{VL} + \left(I_{L\_ref} - \frac{I_L(t)}{I_{d\_base}} \left(\frac{1}{sT_2 + 1} \right) \right) K_{IL} \quad (6.5)$$

$$e_{t2}(t) = \left(\left(\frac{V_d(t)}{V_{d_{base}}} - \frac{V_b(t)}{V_{b_{base}}} \right) \left(\frac{1}{sT_3 + 1} \right) \right) K_{Vdb} + \left(\frac{I_d(t)}{I_{d\_base}} \left(\frac{1}{sT_4 + 1} \right) \right) K_{Id} \quad (6.6)$$

$$e_{Idb}(t) = \left(\frac{I_{dB}(t)}{I_{d_{base}}} \left(\frac{1}{sT_5 + 1} \right) \frac{1}{sD + 1} \right) \quad (6.7)$$

$$e_{tA}(t) = e_1(t) + e_{Idb}(t) \times K_{d1} \quad (6.8)$$

$$e_{tB}(t) = e_{t2}(t) + e_{Idb}(t) \times K_{d2} \quad (6.9)$$

The PID controller is defined with Eq. 6.10 and simulated in Simulink software, as shown in Fig. 6.7d. The error signal is multiplied by K_p, K_i, and K_d coefficients and gathered the output voltage. This equation is designed for PID_A and PID_B input error signals and obtained V_{cA} and V_{cB} output signals as shown in Eqs. 6.11 and 6.12. These output signals are limited and applied in a switched pulse width modulation (SPWM) to generate switching signals:

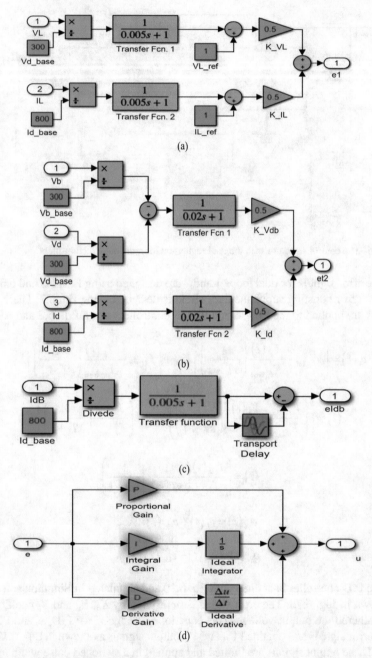

Fig. 6.7 The MATLAB/Simulink models for (**a**) dual loop 1, (**b**) dual loop 2, (**c**) current error boosting supplementary circle, and (**d**) PID controller

$$u(t) = K_p e(t) + K_i \int_0^t e(t) dt + K_d \frac{de(t)}{d(t)} \tag{6.10}$$

$$V_{cA}(t) = K_{p\_A} e_{tA}(t) + K_{i\_A} \int_0^t e_{tA}(t) dt + K_{d\_A} \frac{de_{tA}(t)}{d(t)} \tag{6.11}$$

$$V_{cB}(t) = K_{p\_B} e_{tB}(t) + K_{i\_B} \int_0^t e_{tB}(t) dt + K_{d\_B} \frac{de_{tB}(t)}{d(t)} \tag{6.12}$$

6.4 Digital Simulation Results

The simulation results are obtained in the MATLAB/Simulink software functional model, as shown in Fig. 6.8.

The digital simulation results for scheme A are given in Figs. 6.9, 6.10, 6.11, and 6.12. The load side battery current (I_B), voltage (V_B), and power variations (P_B) for solar irradiation ratio ($S_x = 100\%$) and RL load are seen in Fig. 6.9a. The PV source side current (I_{PV}), voltage (V_{PV}), and power variations (P_{PV}) with time for $S_x = 100\%$ and RL load are seen in Fig. 6.9b. The current and voltage variations in the photovoltaic source side and the battery load side are stable with time for RL load. The load side current (I_B), voltage (V_B), and power variations (P_B) for solar irradiation

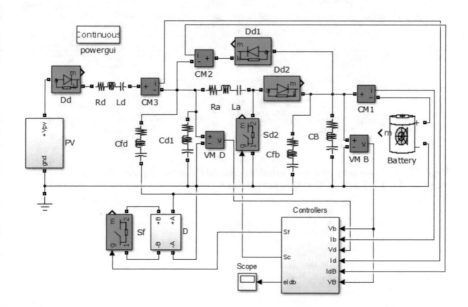

Fig. 6.8 The functional block of the proposed battery charging scheme A in MATLAB/Simulink

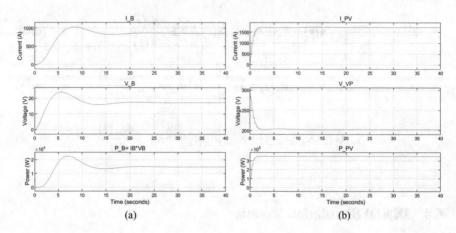

Fig. 6.9 Simulation results of the *RL* load (**a**) PV source (**b**), current, voltage, and power variations for $S_x = 100\%$ and scheme A

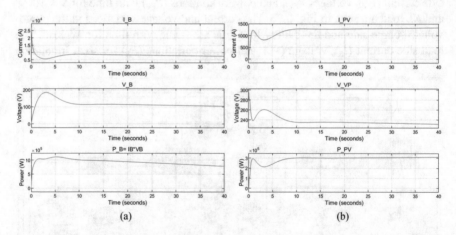

Fig. 6.10 The simulation results of the battery load (**a**) PV source (**b**), current, voltage, and power variations for $S_x = 100\%$, SoC = 90%, and scheme A

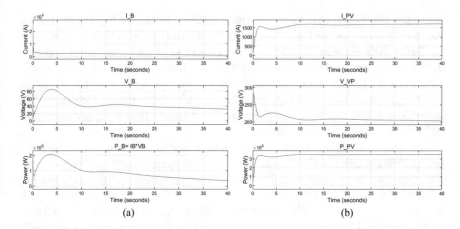

Fig. 6.11 The simulation results of the battery load (**a**) PV source (**b**), current, voltage, and power variations for $S_x = 100\%$, SoC = 10%, and scheme A

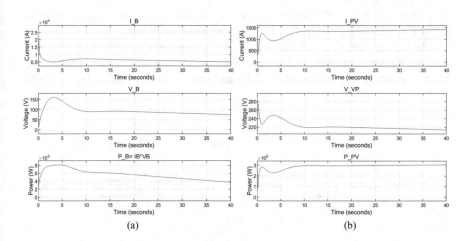

Fig. 6.12 The simulation results of the battery load (**a**) PV source (**b**), current, voltage, and power variations for $S_x = 50\%$, SoC = 50%, and scheme A

ratio ($S_x = 100\%$) and state of charge (SoC = 90%) are seen in Fig. 6.10a. The PV side current (I_{PV}), voltage (V_{PV}), and power variations (P_{PV}) with time for $S_x = 100\%$ and SoC = 90% are seen in Fig. 6.10b. The current and voltage variations in the PV source and battery load side are stable with time. However, there were negative currents from the battery to the converter, and this current flows from the PV to the battery with time. The PV voltage is increased with time to charge the battery. The power flow is constant with time from the PV to battery load time, and more than 90% of PV power charges the battery.

The battery load side current, voltage, and power variations for $S_x = 100\%$ and SoC = 10% are given in Fig. 6.11a. The PV side current, voltage, and power variations with time for $S_x = 100\%$ and SoC = 10% are presented in Fig. 6.11b. The battery load side current, voltage, and power variations for $S_x = 50\%$ and SoC = 50% are seen in Fig. 6.12a. The PV side current, voltage, and power variations with time for $S_x = 50\%$ and SoC = 50% are seen in Fig. 6.12b. The current and voltage variations in the PV side and battery side are stable with time, and the charge of the battery in all states is suitable.

The simulation results for scheme A are given in Figs. 6.13, 6.14, and 6.15. The load side battery current (I_B), voltage (V_B), and power variations (P_B) for solar irradiation ratio ($S_x = 100\%$) and state of charge (SoC = 90%) are seen in Fig. 6.13a. The PV side current (I_{PV}), voltage (V_{PV}), and power variations (P_{PV}) with time for $S_x = 100\%$ and SoC = 90% are seen in Fig. 6.13b. The current and voltage variations in the PV source side and battery load side are stable with time.

The battery load side current, voltage, and power variations for $S_x = 100\%$ and SoC = 10% are seen in Fig. 6.14a for scheme B. The PV side current, voltage, and power variations with time for $S_x = 100\%$ and SoC = 10% are seen in Fig. 6.14b for

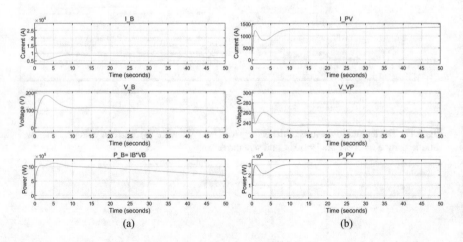

(a)   (b)

Fig. 6.13 The simulation results of the battery load (**a**) PV source (**b**), voltage, current, and power variations for $S_x = 100\%$, SoC = 90%, and scheme B

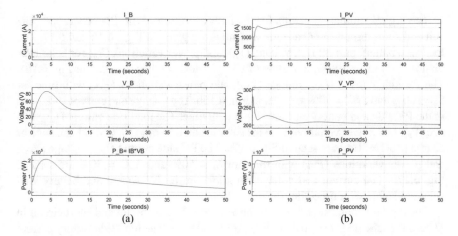

Fig. 6.14 Simulation results of the battery load (**a**) PV source (**b**), current, voltage, and power variations for $S_x = 100\%$, SoC = 10%, and scheme B

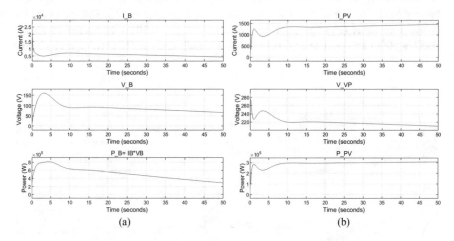

Fig. 6.15 Simulation results of the battery load (**a**) PV source (**b**), current, voltage, and power variations for $S_x = 50\%$, SoC = 50%, and scheme B

scheme B. The battery load side current, voltage, and power variations for $S_x = 50\%$ and SoC = 50% are seen in Fig. 6.15a. The PV side current, voltage, and power variations with time for $S_x = 50\%$ and SoC = 50% are seen in Fig. 6.15b for scheme B. The voltage and current variations in the PV side and battery side are stable with time, and the charge of the battery in all states is suitable.

The digital simulation results for the fault condition of the 10-second open cir-cuit and short circuit in D base for scheme A are given in Figs. 6.16 and 6.17. The load side battery current (I_B), voltage (V_B), and power variations (P_B) for solar irra-diation ratio ($S_x = 100\%$) and state of charge (SoC = 90%) are seen in Fig. 6.16a. The PV side current (I_{PV}), voltage (V_{PV}), and power variations (P_{PV}) with time for $S_x = 100\%$ and SoC = 90% are seen in Fig. 6.16b. The voltage and current variations in the PV side and battery side are stable with time. The load side current (I_B), volt-age (V_B), and power variations (P_B) for solar irradiation ratio ($S_x = 50\%$) and state of charge (SoC = 50%) are seen in Fig. 6.17a. The PV side current (I_{PV}), voltage (V_{PV}), and power variations (P_{PV}) with time for $S_x = 50\%$ and SoC = 50% are seen in Fig. 6.17b. The current and voltage variations in the PV source and battery load sides are stable with time after the open circuit.

The digital simulation results for the ten-second open circuit and short circuit in D base for scheme B are given in Figs. 6.18 and 6.19. The load side battery current (I_B), voltage (V_B), and power variations (P_B) for solar irradiation ratio ($S_x = 100\%$) and state of charge (SoC = 90%) are seen in Fig. 6.18a. The PV side current (I_{PV}), voltage (V_{PV}), and power variations (P_{PV}) with time for $S_x = 100\%$ and SoC = 90% are seen in Fig. 6.18b. The current and voltage variations in the PV source side and battery load side are stable with time. The load side current (I_B), voltage (V_B), and power variations (P_B) for solar irradiation ratio ($S_x = 50\%$) and state of charge (SoC = 50%) are seen in Fig. 6.19a. The PV side current (I_{PV}), voltage (V_{PV}), and

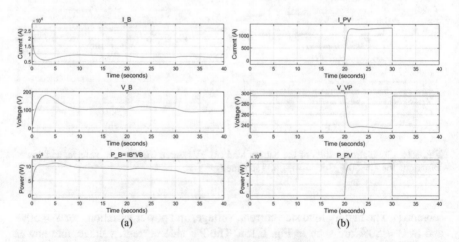

Fig. 6.16 Simulation results of the battery load (**a**) PV source (**b**), current, voltage, and power variations for $S_x = 100\%$ and SoC = 90% during a 10-second open circuit for scheme A

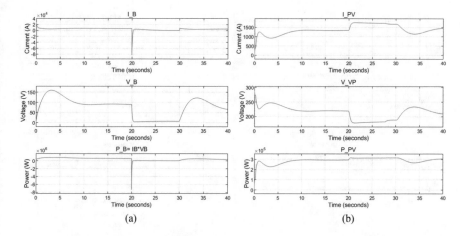

Fig. 6.17 The simulation results of the battery load (**a**) PV source (**b**), current, voltage, and power variations for $S_x = 50\%$ and SoC = 50% during a 10-second short circuit for scheme A

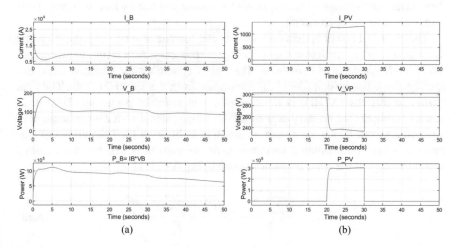

Fig. 6.18 The simulation results of the battery load (**a**), PV source (**b**), current, voltage, and power variations for $S_x = 100\%$ and SoC = 90% during a 10-second open circuit for scheme B

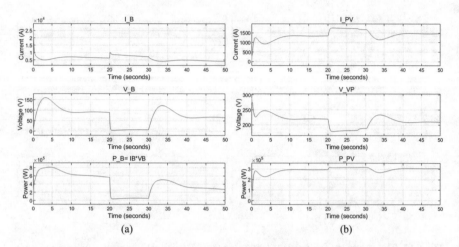

Fig. 6.19 The simulation results of the battery load (**a**), PV source (**b**), current, voltage, and power variations for $S_x = 50\%$ and SoC = 50% during a 10-second short circuit for scheme B

power variations (P_{PV}) with time for $S_x = 50\%$ and SoC = 50% are seen in Fig. 6.19b. The current and voltage variations in the PV source and battery load sides are stable with time after the open circuit.

The electrical power system components can become resonant with the magnetic fields as a result of the harmonics with higher frequency. As a result, these components are frequencies of harmonics. The power system's most common noise frequency range can be considered the harmonics from the 3rd to the 20th. The total harmonic distortion (THD) is represented as an essential parameter. THD is an index to compare the primary element of the voltage signal with the harmonic voltage components, taking voltage as an example as in Eq. 6.13. The number of harmonics is (h), the maximum harmonic order of voltage is (n), and the nominal system voltage at the main frequency is one in Eq. 6.13 [31]:

$$\text{THD} = \frac{\sqrt{\sum_{h=2}^{n} V_h^2}}{V_1} = \frac{\sqrt{V_1^2 + V_2^2 + \ldots + V_n^2}}{V_1} \tag{6.13}$$

THD for the converter input (V_{PV}) and output voltage (V_B) is investigated in Figs. 6.20, 6.21, 6.22, and 6.23 for different conditions. Firstly, for $S_x = 50\%$ and SoC = 50%, V_{PV} and V_B voltage and I_{PV} and I_B current THD analyses for scheme A are given in Fig. 6.20a–d. Secondly, for $S_x = 100\%$ and SoC = 90%, V_{PV} and V_B voltage and I_{PV} and I_B current THD analyses for scheme A are given in Fig. 6.21a–d. The fundamental (50 Hz) harmonics is reduced in the converter output, but THD is slightly increased.

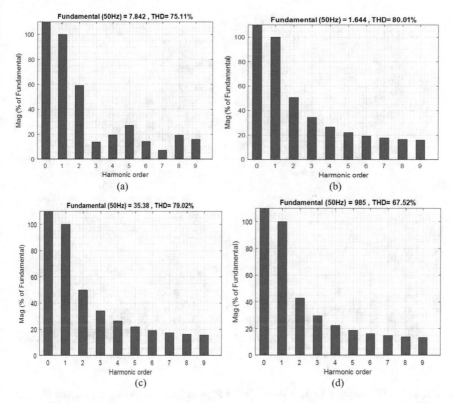

Fig. 6.20 THD graphics for $S_x = 50\%$, SoC = 50%, and (**a**) V_{PV}, (**b**) V_B, (**c**) I_{PV}, and (**d**) I_B in scheme A

For $S_x = 50\%$ and SOC = 50%, V_{PV} and V_B voltage and I_{PV} and I_B current THD analyses for scheme B are given in Fig. 6.22a–d. For $S_x = 100\%$ and SoC = 90%, V_{PV} and V_B voltage and I_{PV} and I_B current THD analyses for scheme B are given in Fig. 6.23a–d. The fundamental (50 Hz) harmonics is decreased in the converter's output, but THD is a little increase.

FFT (fast Fourier transform) is a mathematical algorithm that takes time-domain data and maps it into its frequency domain. Suppose that we have N consecutive sampled values as in Eq. 6.14. We can estimate the Fourier transform $H(f)$ at the discrete values with N numbers of input. The discrete Fourier transform is obtained as in Eq. 6.15. If the number of sampled values is an even power of two, the discrete Fourier transform can be computed in operations with an algorithm called the fast Fourier transform (FFT) [32]:

$$h_k \equiv h(t_k), t_k \equiv k\Delta, k = 0,1,2,\ldots N-1 \tag{6.14}$$

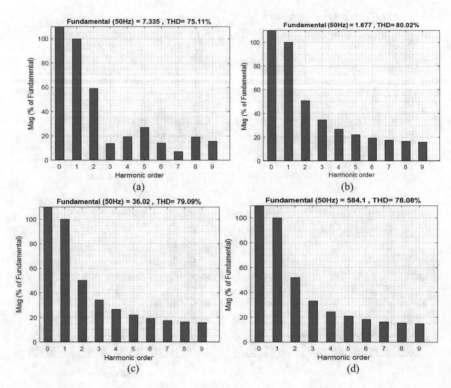

Fig. 6.21 THD graphics for $S_x = 100\%$, SoC = 90%, and (**a**) V_{PV}, (**b**) V_B, (**c**) I_{PV}, and (**d**) I_B in scheme A

$$H(f_n) = \int_{-\infty}^{+\infty} h(t)e^{-2\pi jf_n t} dt \approx \sum_{k=0}^{N-1} h_k e^{-2\pi jf_{nn_1}} \Delta = \Delta \sum_{k=0}^{N-1} h_k e^{-2\pi jk_n / N} \qquad (6.15)$$

The FFT analysis graphics for $S_x = 100\%$, SoC = 90%, V_{PV}, V_B, scheme A, and scheme B are obtained in MATLAB, as shown in Figs. 6.24 and 6.25.

The comparison table for the simulation results of the PV and battery side current voltage and power variations are compared in Table 6.1. The two schemes are compared for different S_x and SoC in this table. The efficiency of the PV-battery system for two schemes and various parameters are compared. The overshoot and response times were compared for the PV and battery sides in other conditions. Also, the short and open circuit test results are given in a comparable form in this table.

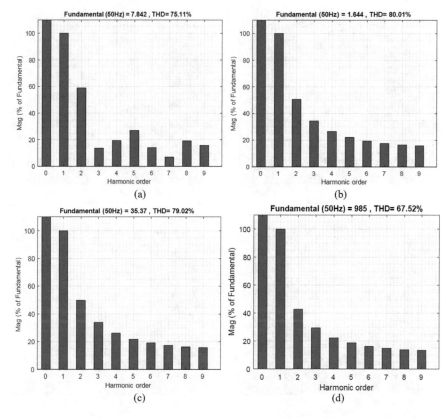

Fig. 6.22 THD graphics for $S_x = 50\%$, SoC $= 50\%$, and (**a**) V_{PV}, (**b**) V_B, (**c**) I_{PV}, and (**d**) I_B in scheme B

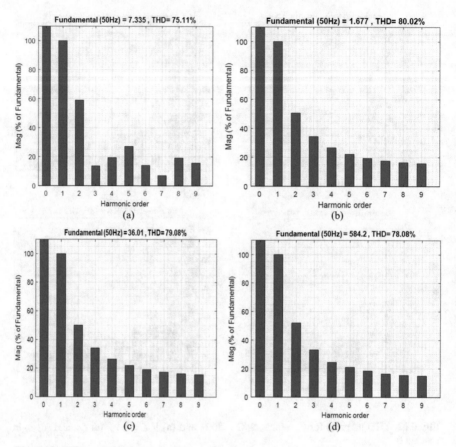

Fig. 6.23 THD graphics for $S_x = 100\%$, SoC = 90%, and (**a**) V_{PV}, (**b**) V_B, (**c**) I_{PV}, and (**d**) I_B in scheme B

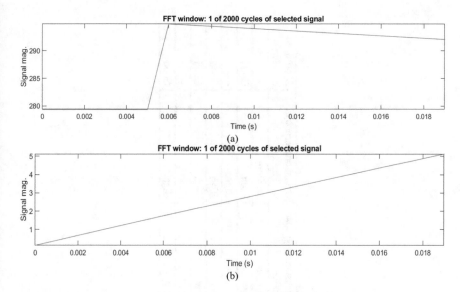

Fig. 6.24 FFT graphics for $S_x = 100\%$, SoC = 90%, and (**a**) V_{PV} and (**b**) V_B in scheme A

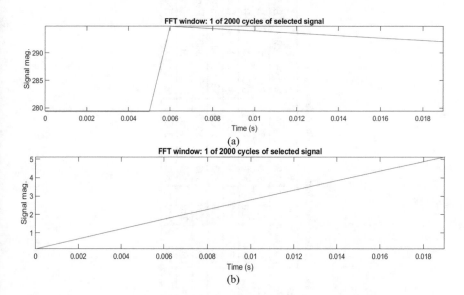

Fig. 6.25 FFT graphics for $S_x = 100\%$, SoC = 90%, and (**a**) V_{PV} and (**b**) V_B in scheme B

Table 6.1 The comparison of the PV and battery side simulation results

Type	S_X (%)	SoC (%)	I_{PV} (A)	V_{PV} (V)	P_{PV} (kW)	I_B (A)	V_B (V)	P_B (kW)	Efficiency	Overshoot response time PV	Overshoot response time battery	Short open circuit test
Scheme A	100	RL	1500	220	330	70	70	5	Low	Medium	Medium	–
	100	90	1300	220	300	8000	100	800	High	Low	High	Good
	100	10	1600	220	330	800	30	24	Low	Low	High	–
	50	50	1400	220	300	5000	80	400	High	Low	High	Good
Scheme B	100	90	1300	220	300	8000	100	800	High	Low	High	Good
	100	10	1600	210	330	1000	30	30	Low	Low	High	–
	50	50	1300	220	300	5000	60	300	High	Low	High	Good

6.5 Conclusions

An efficient PV-powered battery charging scheme that uses two proposed low-cost DC green plug schemes for ripple reduction, efficient energy storage scheme with supercapacitor is validated in this paper. The proposed efficient PV-powered V2H scheme and the basic modules are discussed with a modified multi-regulation decoupled PWM controller. The dual regulation multi-loop controller design and whole scheme basic structures are given in detail. The FFT spectrum analysis for harmonics is investigated for the primary system's PV source input and battery load output. The digital simulation model and results are shown in MATLAB/Simulink software in this paper. The digital simulation results validated an efficient utilization of the green plug schemes A and B in the low-cost reduced ripple-efficient battery charging scheme. The same method can be extended to other microgrids and commercial hybrid source and V2G hybrid multisource smart grid-wind, PV, and fuel cell systems. The controller provides for selective multi-model charging modes or hybrid aggregated mixed modes of battery charging. Other optimal fuzzy, neural network reduced order and incremental controllers with multi-loop regulation can be utilized to ensure fast, efficiently reduced ripple decoupled low-impact energy-efficient charging.

Appendices

Appendix A: Designed GPFC System Parameters

PV model parameters	
Voltage of PV	220 V
Power of PV	300 kW
R_d	0.1 Ω
L_d	50–100 mH
The battery parameters	
Used battery model	Lithium-ion
Nominal battery voltage	240 V
Battery's rated capacity	240 Ah
Boost converter and line parameters	
S_a, S_b, S_c	MOSFETs
C_d	100 mF
R_a	0.01 Ω
L_a	50 mH
C_B	100 F
C_{fD}	6500 uF
C_{fB}	6500 uF

Appendix B: Controller Gain Parameters

Dual-loop controller parameters	
V_{D_base}	300 V
I_{D_base}	800 A
K_{D1}	0.5
K_{D2}	0.25
WM-PID controller parameters	
Proportional (K_P)	0.2
Integrated (K_I)	0.02
Derivative (K_D)	0.01
Switching frequency (f_{sw})	2500 Hz

References

1. F.H. Gandoman, A. Ahmadi, A.M. Sharaf, P. Siano, J. Pou, B. Hredzak, V.G. Agelidis, Review of FACTS technologies and applications for power quality in smart grids with renewable energy systems. Renew. Sust. Energ. Rev. **82**, 502–514 (2018). https://doi.org/10.1016/j.rser.2017.09.062
2. N. Mansouri, A. Lashab, J.M. Guerrero, A. Cherif, Photovoltaic power plants in electrical distribution networks: A review on their impact and solutions. IET Renew. Power Gener. **14**(12), 2114–2125 (2020). https://doi.org/10.1049/iet-rpg.2019.1172
3. A. Slimani, M.N. Tanjaoui, A. Boutadara, L. Saihi, N. Bailek, M.S. Adel, K. Koussa, A PV-active power filter interface scheme for three-phase balanced system. Int. J. Eng. Res. Afr. **46**, 125–145 (2020). https://doi.org/10.4028/www.scientific.net/JERA.46.125
4. M.G. Taul, S. Golestan, X. Wang, P. Davari, F. Blaabjerg, Modeling of converter synchronization stability under grid faults: The general case. IEEE Trans. Emerg. Sel. Topics Power Electron. (Early Access) (2020). https://doi.org/10.1109/JESTPE.2020.3024940
5. S.H. Hanzaei, S.A. Gorji, M. Ektesabi, A scheme-based review of MPPT techniques with respect to input variables including solar irradiance and PV arrays' temperature. IEEE Access **8**, 182229–182239 (2020). https://doi.org/10.1109/ACCESS.2020.3028580
6. A.M. Sharaf, B. Khaki, A FACTS-based switched capacitor compensation scheme for smart grid applications, in *2012 International Symposium on Innovations in Intelligent Systems and Applications*, (Turkey, 2012), pp. 1–5. https://doi.org/10.1109/INISTA.2012.6246986
7. A.M. Sharaf, M.E. Şahin, A novel photovoltaic PV-powered battery charging scheme for electric vehicles, in *2011 International Conference on Energy, Automation and Signal, Bhubaneswar*, (India, 2011), pp. 1–5. https://doi.org/10.1109/ICEAS.2011.6147212
8. M.E. Şahin, H.İ. Okumuş, Physical structure, electrical design, mathematical modeling, and simulation of solar cells and modules. Tur. J. Electromechan. Energy **1**(1), 1–6 (2016)
9. M.E. Şahin, F. Blaabjerg, A hybrid PV-battery/supercapacitor system and a basic active power control proposal in MATLAB/Simulink. Electronics **9**(1), 129 (2020). https://doi.org/10.3390/electronics9010129
10. B. Benlahbib, N. Bouarroudj, S. Mekhilef, T. Abdelkerim, A. Lakhdari, B. Abdelhalim, F. Bouchafaa, Power management and DC link voltage regulation in renewable energy system, in *2019 International Conference on Advanced Electrical Engineering (ICAEE)*, (2019), pp. 1–6. https://doi.org/10.1109/ICAEE47123.2019.9014653

11. A.A. Abdelsalam, H.A. Gabbar, A.M. Sharaf, Performance enhancement of hybrid AC/DC microgrid based D-FACTS. Int. J. Electr. Power Energy Syst. **63**, 382–393 (2014). https://doi. org/10.1016/j.ijepes.2014.06.003

12. F.H. Gandoman, A.M. Sharaf, S.H. Abdel Aleem, F. Jurado, Distributed FACTS stabilisation scheme for efficient utilisation of distributed wind energy systems. Int. Trans. Electr. Energy Syst. **27**(11), e2391 (2017). https://doi.org/10.1002/etep.2391

13. E. Özkop, I.H. Altaş, A.M. Sharaf, A self adjustable FACTS device and controller for distribution systems. Afyon Kocatepe Üniversitesi Fen ve Mühendislik Bilimleri Dergisi **17**(1), 131–137 (2017). https://doi.org/10.5578/fmbd.54015

14. M. Mandour, M. El-shimy, F. Bendary, W.M. Mansour, Damping of power systems oscillations using FACTS power oscillation damper–design and performance analysis. MEPCON Conf. **14**, 15–23 (2014)

15. A. Elgammal, A.M. Sharaf, Multi-objective maximum efficiency self regulating controllers for hybrid PV-FC-diesel-battery electric vehicle drive system. Int. J. Renew. Energy Res. **2**(1), 53–77 (2012)

16. H.A. Gabbar, A. Sharaf, A.M. Othman, A.S. Eldessouky, A.A. Abdelsalam, Intelligent control systems and applications on smart grids, in *New Approaches in Intelligent Control*, (Springer, 2016), pp. 135–163. https://doi.org/10.1007/978-3-319-32168-4_5

17. A.-F. Attia, A.M. Sharaf, F. Selim, A multi-stage fuzzy logic controller for hybrid-AC grid-battery charging drive system. Tur. J. Electromech. Energy **4**(2), 1–12 (2019)

18. A.M. Sharaf, A.A. Elgammal, Novel Green Plug switched filter schemes based on Multi-Objective Genetic algorithm MOGA for single-phase induction motors, in *2010 IEEE Electrical Power & Energy Conference*, (2010), pp. 1–6. https://doi.org/10.1109/EPEC.2010.5697240

19. M.E. Şahin, A.M. Sharaf, A novel efficient PV – Battery powered LED lighting scheme, in *2017 5th International Istanbul Smart Grid and Cities Congress and Fair (ICSG)*, (2017), pp. 90–94. https://doi.org/10.1109/SGCF.2017.7947608

20. A.M. Sharaf, M.E. Şahin, An efficient switched filter compensation used LED lighting PV-battery scheme. J. Circuits, Syst. Comput. **27**(10), 1850156 (2018). https://doi.org/10.1142/S0218126618501566

21. R.W. Erickson, D. Maksimovic, *Chapter 8: Converter Transfer Functions, Fundamentals of Power Electronics* (Springer, 2007)

22. N. Mohan, T.M. Undeland, *Power Electronics: Converters, Applications, and Design* (Wiley, 2007)

23. M.E. Şahin, H.İ. Okumuş, Comparison of different controllers and stability analysis for photovoltaic powered buck-boost DC-DC converter. Electr. Power Compon. Syst. **46**(2), 149–161 (2018). https://doi.org/10.1080/15325008.2018.1436617

24. M.E. Sahin, H.I. Okumus, A fuzzy-logic controlled PV-powered buck-boost DC-DC converter for Battery-Load system, in *2012 International Symposium on Innovations in Intelligent Systems and Applications*, (2012), pp. 1–5. https://doi.org/10.1109/INISTA.2012.6246974

25. A.M. Sharaf, M.E. Şahin, A flexible PV-powered battery-charging scheme for electric vehicles. IETE Tech. Rev. **34**(2), 133–143 (2017). https://doi.org/10.1080/02564602.2016.1155420

26. L.W. Yao et al., Modeling of lithium-ion battery using MATLAB/Simulink, in *IECON 2013-39th Annual Conference of the IEEE Industrial Electronics Society*, (2013). https://doi. org/10.1109/IECON.2013.6699393

27. N. Omar, M.A. Monem, Y. Firouz, J. Salminen, J. Smekens, O. Hegazy, H. Gaulous, G. Mulder, P. Van den Bossche, T. Coosemans, J. Van Mierlo, Lithium iron phosphate-based battery – Assessment of the ageing parameters and development of cycle life model. Appl. Energy **113**, 1575–1585 (2014). https://doi.org/10.1016/j.apenergy.2013.09.003

28. L.H. Saw, K. Somasundaram, Y. Ye, A.A.O. Tay, Electro-thermal analysis of lithium iron phosphate battery for electric vehicles. J. Power Sources **249**, 231–238 (2014). https://doi. org/10.1016/j.jpowsour.2013.10.052

29. O. Tremblay, L.A. Dessaint, Experimental validation of a battery dynamic model for EV applications. World Electr. Veh. J. **3**(1), 1–10 (2009). https://doi.org/10.3390/wevj3020289

30. C. Zhu, X. Li, L. Song, L. Xiang, Development of a theoretically based thermal model for lithium-ion battery pack. J. Power Sources **223**, 155–164 (2013). https://doi.org/10.1016/j.jpowsour.2012.09.035
31. M. Hamdy, M.A. Attia, A.Y. Abdelaziz, S. Kumar, K. Sarita, R.K. Saket, Performance enhancement of STATCOM integrated wind farm for harmonics mitigation using optimization techniques, in *ICT Analysis and Applications*, (Springer, Singapore, 2021), pp. 507–516. https://doi.org/10.1007/978-981-15-8354-4_50
32. P. Duhamel, M. Vetterli, Fast Fourier transforms: A tutorial review and a state of the art. Signal Process. **19**(4), 259–299 (1990). https://doi.org/10.1016/0165-1684(90)90158-U

Chapter 7
Fast Charging for Railways

Hossam A. Gabbar and Yasser Elsayeda

7.1 Introduction

The transportation sector is the second largest source of GHG (greenhouse gas) emission in Canada as it is reported as 25% of the total GHG emissions as shown in details in Fig. 7.1. It shows that GHG emissions from railway, aviation, and marine have shown dramatically an increase from year to year. However, there are still increasing GHG emissions, especially in truck passengers.

Figure 7.2 shows that on average, CMC's Canadian flag fleet can carry 1 ton of cargo for 360 kilometers on 1 liter of fuel, while by rail, it is found to be 247 km and only 45 km for trucks. As a result, the GHG emission would be 31% more for rail and 558% more for trucking to carry the same cargo over the same distance as CMC's fleet thanks to the fuel efficiency, as shown in Fig. 7.3. Therefore, the electrification of rail is essential for reducing GHG emissions.

Railways have been used as the most utilized mean of transportation since decades. Transportation by rail showed a significant investment in the economy in Canada and all over the world. The sector of railway Canada brings $10 billion to the economy as rail transports about 82 million passengers and comes from the goods [1, 2]. Transportation by rail considers 2% of the total energy used by the

The original version of this chapter was revised. The correction to this chapter is available at https://doi.org/10.1007/978-3-031-09500-9_18

H. A. Gabbar (✉)
Faculty of Energy Systems and Nuclear Science, Ontario Tech University (UOIT), Oshawa, ON, Canada

Faculty of Engineering and Applied Science, Ontario Tech University, Oshawa, ON, Canada
e-mail: Hossam.gabbar@uoit.ca

Y. Elsayeda
Faculty of Energy Systems and Nuclear Science, Ontario Tech University (UOIT), Oshawa, ON, Canada

© Springer Nature Switzerland AG 2022, Corrected Publication 2023
H. A. Gabbar, *Fast Charging and Resilient Transportation Infrastructures in Smart Cities*, https://doi.org/10.1007/978-3-031-09500-9_7

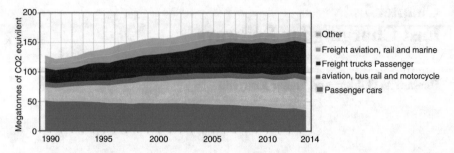

Fig. 7.1 Details of the GHG emission in Canada transportation sector [1]

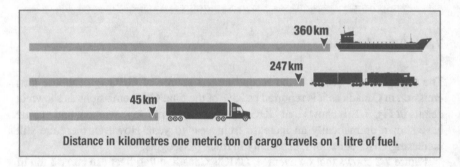

Fig. 7.2 CMC Member fleet fuel efficiency [2]

FIGURE 2. CMC Member Fleet Carbon Emissions

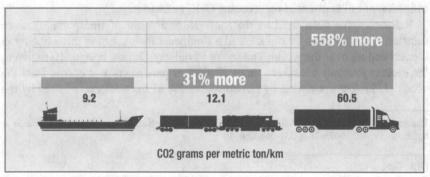

Source: Research and Traffic Group analysis

Fig. 7.3 CMC member fleet carbon emissions [2]

transportation sector, as shown in Fig. 7.4 [3]. Two main categories can be transported by rail: freighted and passengers. Despite transport of freights is very important for national and international, the priority is given to passengers' comfort and requirements. Therefore, rail transportation depends on the grid, so electricity

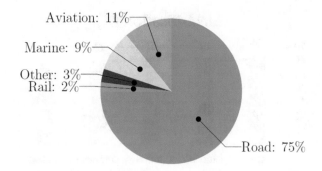

Fig. 7.4 Transportation energy breakdown [3]

outage is a big issue in rail transportation in terms of money for transporting freights and comfort and safety. Therefore, energy storage has a vital role in increasing reliability and reducing loads of the grid. This chapter introduces the infrastructure of fast railway charging and its requirements, topologies, and control strategies.

7.1.1 Chapter Outlines

The outline of this chapter shows that Sect. 7.2 presents the railway electrification infrastructure; Sect. 7.3 presents the voltage standardization for railway electrification; Sect. 7.4 demonstrates the resilient interconnected microgrid (RIMG) for railway; Sect. 7.5 introduces the utility requirements; Sect. 7.6 presents the control system requirements; and Sect. 7.7 presents the conceptual design of interconnected microgrids.

7.2 Railway Electrification Infrastructure

Figure 7.5 charts the electrification of the railway for four countries that are the leaders. Approximately the electrified railway is 31.5% so far worldwide. Japan is the highest with 64%, Russia comes next order with 62%. Then India is with 62%, the European Union with 61%, and China with 46% [3]. An electric railway infrastructure consists of a generation section, transmission part, and distribution network. However, they mainly depend on the electric utility grid [4–6]. Figure 7.6 depicts a typical layout of an electrical power system for supplying an AC electric infrastructure of the railway [4]. In addition, it shows DC electric railway systems.

Moreover, a transformer and AC/DC diode rectifier are included for rectifying the power [4]. It depicts the layout of an AC electrified railway infrastructure. It generally consists of a non-traction transformer that connects the medium- or high-tension voltage of range 10–230 kV AC voltage and 50 and 60 Hz to traction of the power distribution system of 25 kV AC and 50/6 HZ that supplies rail stations.

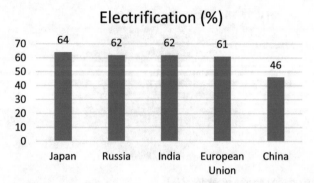

Fig. 7.5 Railway electrification so far in the leader countries [3]

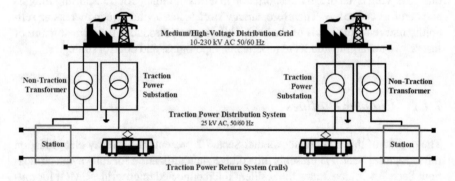

Fig. 7.6 Typical topology of an AC railway electrification

The electrification system can be identified for the railway by three main parameters. The first is the voltage and current, either DC or AC, and frequency. The second is the contacting system of third and fourth rail or overhead transmission lines. Selecting an electrification system depends on the size of investment of energy supply and the cost of maintenance compared with the revenue of both passenger and freight traffic.

7.3   Voltage Standardization for Railway Electrification

Table 7.1 lists the minimum, nominal, and maximum values of voltages for temporary and permanent for railway's most common electrification systems according to BS EN 50163 and IEC 60850 standards [7, 8]. Six electrification systems are the most common, as listed in the first column in Table 7.1. The minimum nonpermanent, the minimum permanent, the nominal, the maximum permanent, and nonpermanent are listed from the second to sixth columns, respectively.

Table 7.1 The voltage standardization railway electrification systems

| Electrification system DC voltage (V) | Voltage | | | | |
	Minimum permanent	Minimum nonpermanent	Typical	Max. permanent	Max. nonpermanent
600	400	400	600	750	850
750	500	500	750	9000	1000
1500	1000	1000	1500	1800	1950
3000	2000	2000	3000	3600	3900
15000 AC, 16.7 Hz	11,000	12,000	15,000	17,250	18,000
25000 AC, 50 Hz (EN 50163) 60 Hz (IEC 60850)	17,500	19,000	25,000	27,500	29,000

7.4 Resilient Interconnected Microgrid (RIMG)

Electric power is supplied via an overhead transmission contact system to the traction power substations (TPSs). There are many TPSs supplied by the energy generated by electrical generation units and distributed by multiple distributed energy resources (DERs) combined with energy storage systems (ESSs). The microgrid ecosystem bidirectionally can send and receive energy to and from the utility grid. The topology of each microgrid consists of AC/DC combined topology. MG is connected via TPSs. In other words, a microgrid is connected with other MGs to exchange energy to diminish the construction of the railway and minimize the usage of the utility grid. The performance of each microgrid is controlled based on a hierarchical topology controller for increasing the system reliability and resiliency of the railway system. In this topology, the IMGs can either store energy in ESSs or provide that energy to another IMG. If there is excess energy, it could be exported to the grid. Based on the time of use, the ESS units can be utilized to save the expensive energy required from the grid at peak time or help during emergencies. For emergencies, plenty of energy is required for covering critical loads like railway stations and systems of communication (Fig. 7.7).

7.5 Requirements of the Utility

Typically, there are generation sector, transmission section, and distribution stations for any utilities. In addition, there should be water utility, telecommunication providers, and gas companies. Some requirements of RIMGs should be satisfied for constructing a valid railway infrastructure. Table 7.2 lists the requirements of the grid operators that should be met for proper, workable electrification system railway and RIMGs. There should be a reliable connection between the electric HV grid and the electrification system of the railway traction. Many of the current electrified

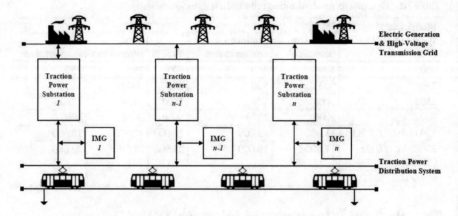

Fig. 7.7 Interconnected Microgrid ecosystem

Table 7.2 The utility requirements for RIMG system

Number	Requirement
1	TPS three-phase, 60 Hz, 230 kV [10]
2	TPS integrated IMGs
3	Load supply: Single-phase, 60 Hz, 25 kV [10]
4	Minimizing the total harmonic distortion (THD) according to IEEE 519-1992 [11]
5	IMG nominal capacity less than 10MVA according to IEEE 1547 [13]
6	Minimizing transmission losses [5]
7	Diminishing the usage of utility grid
8	Switch on and off the IMGs based on the needs
9	Measuring the AC current, voltage, active and reactive power, and power factor of TPSs [12]
10	Supplying the operator with data like the tariff and rate of schedule of the railway
11	The flexibility of the distribution system [14]

networking connections to the utility grid through are either a 115 kV or 230 kV. The operator should specify the limitations of voltage levels of the load to realize the railway's efficiency and security. In addition, American Railway Engineering and Maintenance-of-Way Association (AREMA) determines the voltage boundaries for a railway electrification network by using a 230 kV supplied by the grid as tabulated in Table 7.3 [9].

7.6 The Criteria of the Control System

By analyzing the criteria of the proposed RIMGs, some requirements are highlighted for a hierarchal control system to implement the control. A hierarchical control system is fit immensely for IMGs because the system can be categorized into individual layers to manage the attributes of the ESSs and DERs. These criteria are compulsory for building such a control system. The criteria include every single

Table 7.3 The voltage limits of electrification system based on AREMA 25 kV

Requirement	Voltage (V)
Input voltage	230,000
Upper output voltage limit	27,500
No load output voltage	26,250
The nominal output voltage	25,000
The normal lower voltage limit	20,000
The minimum voltage for emergencies and outage conditions	17,500

Table 7.4 List of criteria of the control system of RIMGs

ID	Criteria
1	Three levels of control structure: Level 1: Suite power converters for DERs and ESSs Level 2: Adapt the DERs' and ESSs' pivot points Level 3: Scanning IMGs' statuses and coordinating energy exchange between all IMGs
2	All reading of DERs and ESSs like SoC and power generation and the demand power should be available to Level 2.
3	Formulating DERs' threshold values by Level 2.
4	The energy at potential points of IMGs should be determined at Level 2 to divert the power toward the railways' loads or the grid [15].
5	The economic and technical calculation for the generation and demand should be calculated.
6	Monitor the threshold values of DERs and ESSs by the controller for assuring the damping of any oscillation [16].
7	DERs must deal with any unexpected imbalances in active power and maintain voltage and frequency at accepted level [15].
8	Response of the controller at Level 1 for any deviation of the threshold values of demand should be within milliseconds and seconds for the Level 2 [17].
9	The IMGs' demand and key performance indicators (KPIs) should be monitored by Levels 2 and 3 [15].
10	IMGs' stability is one of the priorities of the proposed control system to optimize the generation and distribution of power to the loads or the grid [15].
11	Exchanging energy between IMGs is of Level 3 responsibility.

level of the control system. In addition, constraints and assumptions are considered as listed in Table 7.4.

7.7   Design Concepts of Multiple Interconnected Resilient Microgrids

The design concepts of the proposed RIMGs of railway infrastructures [19] are shown in Fig. 7.8. The AC electrification of railway infrastructure was changed to have a single IMG unit at each single TPS. IMGs are connected through the distribution system of the traction power. Practically, TPSs provide energy to some

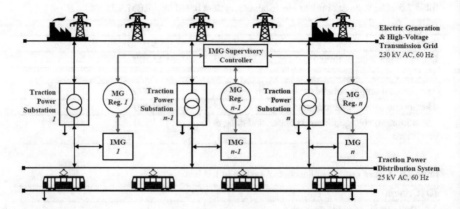

Fig. 7.8 Topology of conceptual design for integrating several RIMGs with TPSs for the railway system

sections of the railway infrastructure thanks to the rolling stock moving in both ways within the section. IMGs will supply the rolling stock with energy between terminal stations. In Fig. 7.8, only two IMGs are considered to show and prove the design concept.

7.8 Design of IMGs

Figure 7.9 shows the design of an IMG that is engaged at the TPS. It shows hybrid AC/DC IMGs for power adapting between IMGs, DERs, and ESS units and diminishes the need for other converters in the IMGs. The DC power generated by DC power sources like the solar PV and battery ESS is inverted to AC power at the AC bus, on which electric grid and railway load are connected.

7.9 Detailed Design of IMGs

Many components are integrated into a hybrid topology with AC and DC buses. For every IMG, there are solar- and WT-based DERs in addition to a battery bank as an ESS unit. The power flow is controlled by a regulation system for control between IMGs to and from the grid and ESSs. In addition, power converters are governed based on local control systems, that is, ESSs and DERs. According to the schematic shown in Fig. 7.10, DERs and ESSs are integrated with the grid to realize the resiliency of power and supplied energy to the railway load. MGs are interconnected through a 25 km feeder, while individual MGs are connected via a 25 km feeder.

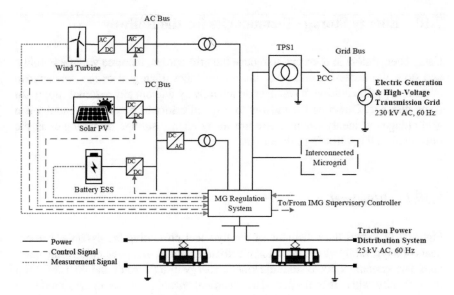

Fig. 7.9 Schematic of a hybrid AC-DC RIMG including all power and energy sources

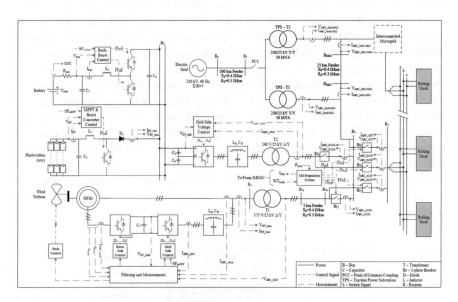

Fig. 7.10 Schematic of the detailed design of a RIMG including all power sources and energy storage

7.10   Energy Storage Technologies for the Railway

Using energy storage is crucial to reduce the grid burden, increase voltage stability, and store the excessive energy of the interconnect microgrid or the regenerative brake energy. In addition, it can operate the railway without any external supply for some distance. Directing the railway to a specific service place is essential when grid outages suddenly occur. There are many energy storage technologies in the railway as in the following subsections.

7.10.1   Flywheel

Figure 7.11 shows the schematic of the flywheel presenting the essential components of a typical flywheel: the rotor, electrical machine, and power electronic interface. The system works to store the kinetic energy into the rotor and then convert it to electricity when it is needed. The electrical machine works in two modes. It works as a motor when charging and as a generator when the flywheel is discharging. On the other hand, battery is one of the most energy storage technologies that are used with the railway. There are many batteries like lead-acid, lithium-ion, nickel-based, sodium-based batteries. In addition, hydrogen fuel cells are another storage technology that can be used with the railway. They are coupled with hydrogen energy storage to compose a regenerative system. Electrolysis of water is used to produce hydrogen to be stored in hydrogen storage which can be converted into electricity thanks to fuel cells. Superconducting magnetic energy storage (SMES) stores energy in a magnetic field by looping DC in a superconducting coil. The stored magnetic energy can be converted to electrical energy by discharging the superconducting coil. Figure 7.12 shows the schematic of the SMEs. Furthermore, this energy storage can be combined to form hybrid energy storage for getting the advantage of the combined energy sources like density, capacity, and lifetime.

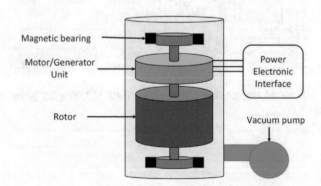

Fig. 7.11   Schematic of flywheel energy storage for railway [18]

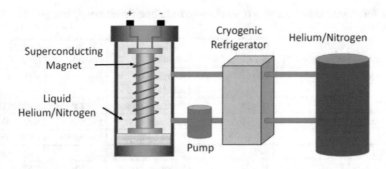

Fig. 7.12 Schematic of SME energy storage for railway [18]

Table 7.5 A comparison between the energy and power density of different energy storage technologies for railways

Technology	Gravimetric energy density (Wh/kg)	Gravimetric power density (W/kg)	Volumetric energy density (Wh/L)	Volumetric power density (W/L)
Flywheel	5–100	400–1500	20–80	1000–2000
EDLC	5–15	5000–10,000	10–30	>100,000
Lead-acid	30–50	75–300	50–90	10–400
Ni-Cd	50–75	150–300	60–150	75–700
Ni-MH	54–120	200–1200	190–490	500–3000
NaS	150–240	150–230	150–250	140–180
ZEBRA	100–120	150–200	150–180	220–300
Li-ion	150–250	500–2000	400–650	1500–10,000
VRFB	10–30	166	25–35	<2
HFC	800–10,000	5–800	500–3000	>500
SMES	0.5–5	500–2000	0.2–2.5	1000–4000

7.10.2   ESS in Railway Systems

In this section, the different types of energy storage technologies are compared to show the pros and cons of each and the feasibility of being used with the railway. The comparative study includes energy density, power density, cycle efficiency, self-discharge, storage duration, service life, capital cost, and environmental impacts. It shows that the flywheel and SMEs are of the most significant power density. Some battery technologies are higher in energy density (Table 7.5).

Table 7.6 lists the other characteristics of the different energy storage technologies for the railway. It shows that flywheel, SMEs, and lithium-ion batteries have the highest cycle efficiency. In addition, the flywheel has the fastest discharge while the batteries have a longer lasting time. For the lifestyle, the SMEs have the highest cycle life while most batteries have fewer cycles.

Table 7.6 A comparison between other characteristics of different energy storage technologies for railway

Technology	Cycle efficiency (%)	Daily self-discharge (%)	Typical storage duration	Lifetime (years)	Cycle life (cycles)
Flywheel	90–95	100	Seconds–minutes	20	>21,000
EDLC	90–97	10–20	Seconds–hours	10–30	>100,000
Lead-acid	80–90	0.05–0.3	Minutes–days	5–15	500–2000
Ni-Cd	60–83	0.2–0.6	Minutes–days	15–20	1500–3000
Ni-MH	65–70	1–2	Minutes–days	15–20	1500–3000
NaS	75–90	15–20	Seconds–hours	10–20	2000–4500
ZEBRA	90	10–15	Minutes–hours	10–20	>2500
Li-ion	90–98	0.1–0.3	Minutes–days	8–15	1000–10,000
VRFB	75–80	0–10	Hours–months	10–20	>16,000
HFC	20–50	Almost zero	Hours–days	5–15	>1000
SMES	94–97	10–15	Minutes–hours	>20	>100,000

Acknowledgments The authors would like to thank the Smart Energy Systems Laboratory's (SESL) support at Ontario Tech University.

References

1. Rail Transportation – Transport Canada. [Online]. Available https://www.tc.gc.ca/eng/policy/anre-menu-3020.htm
2. R.A. of Canada, Railways 101, 2016. [Online]. Available https://www.railcan.ca/railways-101/
3. Railway Handbook 2017, International Energy Agency & International Union of Railways, Paris, Technical Report 2017. [Online]. Available http://uic.org/IMG/pdf/handbookiea-UIC2017web3.pdf
4. A. Gonzalez-Gil, R. Palacin, P. Batty, J.P. Powell, A systems approach to reduce urban rail energy consumption. Energy Convers. Manag. **80**, 509–524 (April 2014)
5. F. Schmid, C. Goodman, Overview of electric railway systems, in *IET Conference Proceedings*, vol. 44, (The Institution of Engineering & Technology, 2014), pp. 1–15
6. A.J. Lopez-Lfiopez, R.R. Pecharromfian, A. Fernfiandez-Cardador, A.P. Cucala, Assessment of energy-saving techniques in direct-current-electrified mass transit systems. Transp. Res. Part C Emerg. Technol. **38**, 85–100 (October 2014)
7. EN 50163: Railway Applications. Supply Voltages of Traction Systems (2007)
8. IEC 60850, *Railway Applications – Supply Voltages of Traction Systems*, 3rd edn. (2007)
9. S. Ma, B. Chen, Z. Wang, Resilience enhancement strategy for distribution systems under extreme weather events. IEEE Trans. Smart Grid **9**(2), 1442–1451 (March 2018)
10. GO Electrification Study Final Report, Delcan Arup, Toronto, Technical Report 2010. [Online]. Available http://www.gotransit.com/electrification/en/projecthistory/docs/ElectricificationStudyFinalReport.pdf
11. Y. Han, P. Shen, X. Zhao, J.M. Guerrero, Control strategies for islanded microgrid using enhanced hierarchical control structure with multiple current – Loop damping schemes. IEEE Trans. Smart Grid **8**(3), 1–15 (May 2015)
12. D. Polhill, *Key Train Requirements*, Technical Report 4 (Association of Train Operating Conditions, London, 2016)

13. H. Farzin, M. Fotuhi-Firuzabad, M. Moeini-Aghtaie, Enhancing power system resilience through hierarchical outage management in multi-microgrids. IEEE Trans. Smart Grid 7(6), 2869–2879 (November 2016)

14. G.A. Bakke, *The Grid: The Fraying Wires Between Americans and our Energy Future* (Bloomsbury USA, New York, 2016)

15. E. Unamuno, J.A. Barrena, Hybrid ac/dc microgrids – Part II: Review and classification of control strategies. Renew. Sust. Energ. Rev. **52**, 1123–1134 (December 2015)

16. S.S. Thale, R.G. Wandhare, V. Agarwal, A novel reconfigurable microgrid architecture with renewable energy sources and storage. IEEE Trans. Ind. Appl. **51**(2), 1805–1816 (March 2015)

17. T. Mostyn, B. Hredzak, V.G. Agelidis, Control strategies for microgrids with distributed energy storage systems: An overview. IEEE Trans. Smart Grid **9**(4), 3652–3666 (July 2016)

18. X. Liu, K. Li, Energy storage devices in electrified railway systems: A review. Transp. Saf. Environ. **2**(3), 183–201 (2020)

19. H.A. Gabbar, T. Egan, A.M. Othman, R. Milman, Hierarchical control of resilient interconnected microgrids for mass transit systems. J. IET Electr. Syst. Transp. (2020). https://doi.org/10.1049/iet-est.2019.0145

Chapter 8
Hybrid Charging Stations

Hossam A. Gabbar

Acronyms

VMT	vehicle miles traveled
TES	thermal energy storage
ICV	internal combustion vehicle
EV	electric vehicle
DV	diesel vehicle
FCV	fuel cell vehicle
MEG	micro energy grid

Nomenclature

$EV_i(i,j)$	Charging current of EV (i) at trip (j) (A)
$EV_t(i,j)$	Charging time of EV (i) at trip (j) (h)
$EV_{cm}(i,j)$	Charging mode of EV (i) ("W," "D")
$ICV_{ga}(i,j)$	Fueling gas amount of ICV (i) at trip (j) (liter)
$ICV_{em}(i,j)$ (kg)	GHG emissions from ICV (i) at trip (j)
$DV_{ga}(i,j)$	Fueling diesel amount of DV (i) at trip (j)(liter)
$FCV_{ga}(i,j)$	Fueling H2 amount of FCV (i) at trip (i) (liter)
$PV_i(i,j)$	Current of PV (i) at trip (j) (A)
$PV_v(i,j)$	Voltage of PV during EV (i) at trip (j) (V)
$CHP_{te}(i,j)$	Thermal efficiency of CHP during EV (i) at trip (j) (%)

This chapter is contributed by Hossam A. Gabbar.

H. A. Gabbar (✉)
Energy Systems and Nuclear Science, University of Ontario Institute of Technology,
Oshawa, ON, Canada
e-mail: Hossam.gabbar@uoit.ca

© Springer Nature Switzerland AG 2022
H. A. Gabbar, *Fast Charging and Resilient Transportation Infrastructures in Smart Cities*, https://doi.org/10.1007/978-3-031-09500-9_8

$CHP_{pe}(i,j)$	Power efficiency of CHP during EV (i) at trip (j) (%)
$CHP_v(i,j)$	Output voltage of CHP during EV (i) at trip (j) (V)
$CHP_i(i,j)$	Output current of CHP during EV (i) at trip (j) (A)
$CHP_t(i,j)$	Output temperature of CHP during EV (i) at trip (j) (C)
$WT_{pe}(i,j)$	Power efficiency of wind turbine during EV (i) at trip (j) (%)
$WT_{op}(i,j)$	Output power of wind turbine during EV (i) at trip (j) (W)
$WTE_{op}(i,j)$	Output power of waste-to-energy during EV (i) at trip (j) (W)
$WTE_{pe}(i,j)$	Power efficiency of waste-to-energy during EV (i) at trip (j) (%)
$ELT_{oh2e}(i,j)$	H2 production efficiency of electrolyzer during EV (i) at trip (j) (%)
$ELT_{ip}(i,j)$	Input power of electrolyzer during EV (i) at trip (j) (W)
$FC_{op}(i,j)$	Output power of fuel cell during EV (i) at trip (j) (W)
$FC_{pe}(i,j)$	Power efficiency of fuel cell during EV (i) at trip (j) (%)
$GRD_{ps}(i,j)$	Power supplied by the grid during EV (i) at trip (j) (W)
T_{st}	Thermal stored in thermal storage system (Joules)
T_{uss}	Thermal used in stations (EV, ICV, FCV) (Joules)
T_{ge}	Thermal generated from MEG (Joules)
P_{sgm}	Power supplied by the grid to MEG (W)
P_{rmg}	Power received from MEG to the grid (W)
P_{ses}	Power supplied by MEG to EV station (W)
P_{sgs}	Power supplied from MEG to ICV station (W)
P_{shs}	Power supplied from MEG to FCV station (W)
P_{cws}	Power used for wireless charging of EVs (W)
P_{cwd}	Power used for wired charging of EVs (W)
G_{su}	Gas supplied from gas transmission to station (liter)
G_{gm}	Gas generated by MEG to ICV station (liter)
G_{sus}	Total gas supplied to ICV station (liter)
G_{us}	Gas used at station to fuel ICV (liter)
D_{us}	Diesel supplied from diesel transmission to station (liter)
D_{gm}	Diesel generated by MEG to DV stations (liter)
D_{sum}	Total diesel supplied to DV station (liter)
D_{us}	Diesel used at station to fuel DV (liter)
W_{su}	Waste collected to MEG for WTE (kg)
H_{su}	Hydrogen supplied from hydrogen supply chain to FCV station (liter)
H_{ge}	Hydrogen generated by MEG to FCV station (liter)
H_{us}	Hydrogen used at station for FCV (liter)
EV_{mwt}	EV maximum waiting time till start charging at EV station (h)
ICV_{mwt}	ICV maximum waiting time till start refueling at ICV station (h)
FCV_{mwt}	FCV maximum waiting time till start refueling at FCV station (h)
SoC_{min}^{EV}	EV minimum SoC after charging (%)
SoC_{min}^{ICV}	ICV minimum SoC after charging (%)
SoC_{min}^{FCV}	FCV minimum SoC after charging (%)
EV_{msc}	Maximum number of simultaneous EV charging at EV station
EV_{msf}	Maximum number of simultaneous ICV fueling at ICV station
EV_{msf}	Maximum number of simultaneous FCV fueling at FCV station

8.1 Introduction

In Europe, the transportation sector represents more than 30% of energy consumption. The corresponding greenhouse gas emission percentage in 2015 was 23.5%, where oil represented 94% of the energy supply. Hydrogen is introduced to the transportation sector to reduce greenhouse gas emissions. A fuel cell vehicle (FCV) is an electric vehicle (EV) with a fuel cell. The integration of fuel cells with battery or ultracapacitor in EV is called fuel cell hybrid electric vehicle (FCHEV). FCHEV will provide EV with more energy density and better dynamic response. The energy management optimization of FCV or FCHEV will lead to enhanced deployment of hydrogen in transportation and will improve the associated economic values [1]. Hybrid charging stations are introduced to charge both EV and fuel cell vehicles. In the hybrid charging station, PV, battery, and electrolyzer with fuel cell configuration are used to charge EV and fuel cell vehicles [2]. The model was evaluated with HOMER to demonstrate performance optimization with an energy management system based on design and control parameters. The integration of the fuel cell charging station with the grid will enable V2G (vehicle-to-grid) capabilities during the charging-discharging of fuel cell vehicles [3]. The optimized operation improved grid performance while increasing the penetration of renewable energy.

To support the penetration of fuel cell vehicles, it is essential to have effective hydrogen production and supply chain in place. There are several scenarios to achieve hydrogen production. There are limitations of using fossil fuels to produce hydrogen, such as greenhouse gas emissions. Hydrogen production from water, biomass, and thermochemical conversion are promising methods, while renewable energy can offer competitive advantages. The different hydrogen production methods are analyzed using indexes of energy, exergy, environment, social, and economy depending on the conversion quantities [4]. Lifecycle assessment is used to evaluate EV and FCV while considering well-to-tank and tank-to-wheel analysis approaches, where greenhouse gases are also reduced [5].

Renewable natural gas (RNG) and compressed renewable natural gas (CRNG) are potential fuel options for heavy-duty trucks with improved performance compared to diesel, with considerations of different traffic models [6]. Hydrogen production from renewable natural gas is also analyzed in a case study in Shanghai with a hydrogen fueling station, along with 9 other hydrogen production methods and 6 possible transportation solutions, which formed 12 hydrogen supply options [7]. Onsite hydrogen production with wind turbine and grid support within charging stations is evaluated based on cooperative and bargaining models with energy trading [8].

Waste-to-energy facilities convert waste with gasification or pyrolysis into clean oil and electricity. The interactions between waste-to-energy and transportation are represented by transporting waste in community applications. Also, integrating waste-to-energy facilities with charging stations will add another dimension of utilizing generated clean energy from the local community to support charging transportation infrastructures [9]. The links among energy, transportation, and waste treatment are modeled within supply chains while considering lifecycle cost analysis [10]. The study indicated the benefits of establishing the coupling between waste and transportation.

Combined heat and power (CHP) system is integrated within a microgrid where renewable energy sources are scheduled in view of electric and thermal loads [11]. The integrated CHP with microgrid will offer balanced electricity capacity to meet charging loads.

Given the different hybrid energy systems, the hybrid design of transportation charging station seems promising. Possible designs of the hybrid charging station are explained in the following section.

8.2 Hybrid Charging Station

The existing gas fueling stations are well established to provide complete coverage of refueling to internal combustion (IC) vehicles (ICVs). The distribution of gas stations is well planned to cover different car trips at any road. ICV can find the closest gas station regardless of the status of ICV gas tank. Similarly, diesel vehicles (DVs) can also find diesel refueling stations for any given tank status on any road. Similarly, fuel cell vehicles (FCVs) will require hydrogen (H2) production and supply chain infrastructure to reach every FCV on the road and integrate hydrogen refueling stations to charge FCVs.

The transition toward electric vehicle (EV) requires establishing adequate charging stations at all roads to support different EV ranges, trips, and battery SoCs.

The upgrade of the existing gas fueling station is strategic where it will reduce possible capital costs to establish EV charging infrastructures. The share of land, facilities, utility services, and energy and power connections will dramatically reduce the costs of the expansion of charging infrastructures.

The possible design of a hybrid charging station is shown in Fig. 8.1. It shows integrated charging functions for EV, ICV, DV, and FCV. Integration with thermal energy storage (TES) is linked to thermal lines. Gas supply lines are linked to MEG and gas stations.

The hybrid design of the charging station will include a micro energy grid (MEG) to complement the grid. Possible V2G capabilities will allow electricity back to the grid with enhanced grid performance. EV charging unit is used to charge EV via wireless or wired links. Gas charging unit will charge ICVs, while diesel charging unit will charge DVs. Hydrogen refueling unit will charge H2 into FCVs. The MEG includes PV, wind turbine (WT), battery (BT), fuel cell (FC), combined heat and power (CHP) unit, and waste-to-energy (WTE) unit. The electrolyzer is integrated to generate hydrogen, which is used by fuel cell within the MEG, or to charge FCV in the hydrogen refueling unit. The proposed MEG is designed based on different control strategies and optimized for charging loads of EV, ICV, DV, and FCV, while maximizing grid performance.

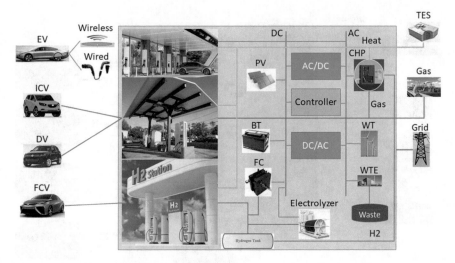

Fig. 8.1 Possible design of hybrid charging station

The overall MEG charging station will have performance measures related to DC and AC buses, H2 line, thermal line, and waste line. Each unit's design and control parameters are defined and associated with performance measures. Overall performance is optimized to support each party, namely, station owner, vehicle driver, power utility, and waste collection municipality.

8.3   Operation of Hybrid Charging Station

The possible scenarios to operate the hybrid charging station for electric vehicle (EV), internal combustion vehicle (ICV), diesel vehicle (DV), and fuel cell vehicle (FCV) can be explained using the charging process shown in Fig. 8.2.

The charging process and interactions among energy supplies and charging stations can be modeled with the considerations of possible scenarios, as explained in Table 8.1.

In these scenarios, gas generation options within MEG are excluded to ensure clean charging infrastructures. Lifecycle analysis (LCA) is used to evaluate possible options with the value of emissions to provide the accurate price of energy supply and conversion to reach charging stations.

The sizing of different components within the integrated charging stations and MEG will be evaluated based on each station's incoming vehicles as load profiles. Optimization techniques based on AI (artificial intelligence) algorithms will be used to achieve overall optimum performance. The optimization of performance measures will be achieved to support all parties in the model: the driver of EV/ICV/DV/FCV, owner/operator of charging stations, power utility company, gas utility company, hydrogen utility company, thermal storage owners, and municipalities.

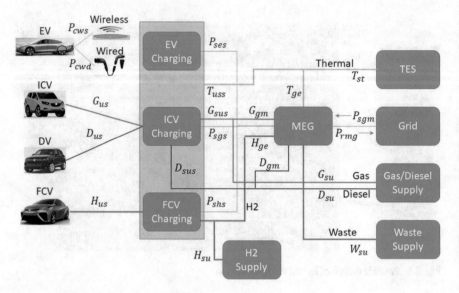

Fig. 8.2 Operation of hybrid charging station

Table 8.1 Scenario modeling for the hybrid charging station

Charging unit	Supply	Resource	Factors
EV charging power	Grid		Grid condition Power quality Electricity price
	MEG	PV WT FC + H2 CHP + Gas BT WTE conversion	PV generation capacity WT generation capacity H2 price + LCA Gas price + LCA BT size Waste amount + LCA
ICV charging	Gas supply		Gas price
	MEG	WTE conversion	Input waste
DV charging	Diesel supply		
	MEG	WTE conversion	Input waste
FCV charging	H2 supply		H2 price
	MEG	Electrolyzer	PV generation capacity WT generation capacity CHP + gas price Waste amount + LCA Electricity price + LCA

Performance measures or key performance indicators (KPIs) for each component and overall system are defined, as shown in Table 8.2.

A collaborative approach could be used to ensure reduced GHG emissions and enhance overall performance [12]. For example, power utility companies will try to maximize electricity sales, while gas utility companies will try to maximize gas

Table 8.2 Performance measures (KPIs) of the hybrid charging station

Component	KPI	Parameters
EV charging unit	Average wireless charging time per EV (h) Average wired charging time per EV (h) Average charging rate per EV (diff-SoC/h)	Number of wireless EV charged per day Number of wired EV charged per day Wireless charging time per EV (h) Wired charging time per EV (h) SoC of EV before wireless charge (%) SoC of EV before wired charge (%) SoC of EV after wireless charge (%) SoC of EV after wired charge (%) Wireless charge power rate (W) wired charge power rate (W)
ICV charging unit	Average fueling time per ICV (h) Average fueling rate per ICV (diff-SoC/h)	Number of ICV fueled per day Fueling time per ICV (h) SoC of ICV before fueling (%) SoC of ICV after fueling (%) Fueling rate (liter)
DV charging unit	Average wireless charging time per EV (h) Average wired charging time per EV (h) Average charging rate per EV (diff-SoC/h)	Number of DV fueled per day Fueling time per DV (h) SoC of DV before fueling (%) SoC of DV after fueling (%) Fueling rate (liter)
FCV charging unit	Average wireless charging time per EV (h) Average wired charging time per EV (h) Average charging rate per EV (diff-SOC/h)	Number of FCV fueled per day Fueling time per FCV (h) SoC of FCV before fueling (%) SoC of FCV after fueling (%) Fueling rate (kg)
EV charging station	Energy efficiency of wireless charging: Total wireless energy charged/total energy supplied for wireless charging unit (%) Energy efficiency of wired charging: Total wired energy charged/total energy supplied for wired charging unit (%)	Number of EV charged per day Charging time per EV (h) SoC of EV before charge (%) SoC of EV after charge (%) Charge power rate (W)

sales. Similarly, hydrogen suppliers will try to maximize hydrogen sales. The trade among the different players will be based on negotiation. Game theory is used to resolve the energy trade based on identified factors and performance measures.

8.4   Data Analysis of Hybrid Charging Station

8.4.1   EV Charging Station Data

Number of wireless EV charged per day: 25
Number of wired EV charged per day: 40
Wireless charging time per EV (h): in minutes
Sample data as below:

367 257 514 459 432 342 236 305 426 320 323 336 515 517 343 420 244 492 235 246 215 564 488 502 229

Wired charging time per EV (h) in minutes
Sample data as below:

199 114 97 345 131 231 105 150 240 151 90 205 171 304 186 98 89 202 180 328 330 136 143 167 230 185 294 144 191 147 326 201 159 134 347 158 213 102 203 220

SoC of EV before wireless charge (%)
Sample data as below:

25 9 58 20 15 23 12 98 26 8 44 83 11 31 51 47 38 100 14 73 54 24 59 43 74

SoC of EV before wired charge (%)
Sample data as below:

83 44 33 16 53 92 8 42 34 90 50 79 37 32 35 43 89 18 21 14 62 68 25 9 17 88 67 70 49 81 56 24 69 58 72 98 80 91 41 94

SoC of EV after wireless charge (%)
Sample data as below:

95 83 84 81 97 89 87 94 88 83 97 81 80 85 94 85 100 93 84 82 81 88 91 83 100

SoC of EV after wired charge (%)
Sample data as below:
96 99 88 91 87 91 95 94 84 85 100 81 95 81 84 95 92 100 82 88 97 86 91 94 88 93 80 87 83 92 84 85 96 100 91 95 98 87 87 82
Wireless charge power rate (W) and wired charge power rate (W) (Table 8.3)

Table 8.3 Power rate for EV charging

Levels	Maximum power rating (kW)
SAE Standard	
AC charging	133
DC charging	
Level 1	350
Level 2	1200
CHAdeMO	
DC charging	
Level 1	62.5
Level 2	400
Oppcharge (iec) Standard	
DC charging	
Level 1	150
Level 2	300 and 450
DC fast charging [13]	
Stand-alone: Comprised of a single unit, stand-alone charging stations	50–250 kW
Split: Charging stations with split architecture come with two main components—a user unit and a power unit	175–350 kW

8.4.2   Gas Refueling Station Data

There are 96 distinct brands of gasoline that could be used for transportation and industrial applications. The price-controlled gas stations represent 21% of all gas stations by Canada's integrated refiner marketers. It was 32% in 2004. Price-controlled gas stations by individual companies represents 79% of all gas stations. This includes companies not involved in the refining process [14] (Tables 8.4, 8.5, and 8.6).

Number of ICV fueled per day: 80
Fueling time per ICV (h): 0.2 (average)
SoC of ICV before fueling (%): 5% (average)
SoC of ICV after fueling (%): 95% (average)
Fueling rate (liter): 55 (average)

Considering 115,000 gas stations in the United States, around 11,000,000 vehicles visit gas stations per day. This will give around 95 vehicles to visit each gas station per day (on average).

8.4.3   FCV Refueling Station Data

Worldwide, by the end of 2020, there will be **31,225 passengers of FCVs** powered with hydrogen (Table 8.7).

Table 8.4 Gas station fueling amounts [15]

Geography	Canada				
	2016	2017	2018	2019	2020
Type of fuel sales	Liters				
Net sales of gasoline 3	42,781,270	43,652,612	43,532,052	43,335,319	36,894,411
Gross sales of gasoline 4	45,086,441	44,966,853	44,796,754	44,806,199	38,597,569
Net sales of diesel oil 3	17,019,229	18,056,530	18,257,318	17,836,573	16,222,705
Net sales of liquefied petroleum gas 3	530,171	343,405	441,549	435,879	503,848

Table 8.5 Gas consumption with vehicle types [15]

Type of vehicle body	Type of fuel	2005	2006	2007	2008	2009
		Liters				
Total, all vehicles' body types	Gasoline	29,677.8E	31,111.3D	31,624.8C	30,312.3E	31,460.5E
	Diesel	10,135.4B	10,075.4B	11,068.9A	10,673.8A	9898.0A
Car	Gasoline	13,715.7E	13,087.8D	12,658.9D	12,708.4E	14,324.7E
	Diesel	F	F	F	F	200.4E
Station wagon	Gasoline	F	F	F	949.8C	1105.7E
	Diesel	F	F	F	F	F
Van	Gasoline	6120.0E	6289.0E	6379.4E	5124.6E	4923.4E
	Diesel	F	F	F	119.4E	158.5E
Sport utility vehicle	Gasoline	F	3227.2E	4409.8E	4333.9E	5163.8B
	Diesel	F	F	F	F	F
Pickup truck	Gasoline	6080.1E	7283.3C	7467.7C	6954.7A	5653.2E
	Diesel	1097.9E	1196.9E	1236.1E	927.7B	1002.1B
Straight truck	Gasoline	F	217.4E	208.3E	184.8E	247.7E
	Diesel	2280.8B	2151.1B	2289.7B	2455.2B	2233.8B
Tractor-trailer	Gasoline	...	...	...	...	...
	Diesel	6342.5A	6366.5A	7222.0A	6875.6A	6195.3A
Bus	Gasoline	F	F	F	F	F
	Diesel	F	F	F	F	F
Other body types	Gasoline	F	F	F	29.4E	F
	Diesel	F	F	F	F	F

Symbol legend:
… Not applicable, E Use with caution, F Too unreliable to be published, A Data quality: excellent, B Data quality: very good, C Data quality: good, D Data quality: acceptable

Number of FCV fueled per day in the United States: 2048
Fueling time per FCV (h): 0.08–0.1
SoC of FCV before fueling (%): 5–15
SoC of FCV after fueling (%): 85–100
Fueling rate (liter/h): 122/0.08 = 1525 liter/h
FCV tank: 122 liters

Table 8.6 Average annual fuel use per vehicle type [16]

Vehicle type	MPG gasoline	MPG diesel	VMT	Annual fuel use (GGE)	Source
Transit bus	3.3	**3.7**	43,647	13,329	A
Class 8 truck	5.3	**6.0**	62,751	11,818	B
Refuse truck	2.5	**2.8**	25,000	10,089	C
Paratransit shuttle	7.1	**8.0**	29,429	4157	A
Delivery truck	6.5	**7.4**	12,435	1899	B
School bus	6.2	**7.0**	12,000	1937	D
Light truck/van	**17.5**	19.8	11,543	660	B
Car	**24.2**	27.3	11,467	474	B
Motorcycle	**44.0**	49.7	2312	53	B

Table 8.7 Fuel cell vehicle data as of January 1, 2022 [17]

FCVs—Fuel cell cars sold and leased in United State	**12,283**
FCBs—Fuel cell buses in operation in California	**48**
Fuel cell buses in development in California	58
Hydrogen stations available in California	**49**
Retail hydrogen stations in construction in California	10
Retail hydrogen stations in permitting in California	27
Retail hydrogen stations proposed in California	16
Retail hydrogen stations funded, but not in development in California	71
Total retail hydrogen stations in development in California	**124**
Retail truck hydrogen stations in construction in California	4
Retail truck hydrogen stations funded, but not in development in California	5

8.5   Optimization of Hybrid Station Operation

8.5.1   Objective Functions

EV Station

>   Maximize the total number of charged EV per day
>   Maximize total power to charge EVs per day
>   Minimize wait time to start charging EV

ICV Station

>   Maximize the total number of fueled ICV per day
>   Maximize total fuel for ICV per day
>   Minimize wait time to start fueling EV

FCV Station

> Maximize the total number of fueled FCV per day
> Maximize total fuel for FCV per day
> Minimize wait time to start fueling FCV

Power Grid

> Maximize total power sent back to the grid per day
> Minimize total power received from the grid per day

Gas Supply

> Maximize total gas/diesel received from the gas/diesel supply per day

MEG

> Maximize power generated from MEG per day
> Maximize hydrogen generated from MEG per day

8.5.2 Constraints

EV waiting time at charging station till start refueling at EV station $< EV_{mwt.}$
ICV waiting time at refueling station till start refueling at ICV station $< ICV_{mwt}$
FCV waiting time at refueling station till start refueling at FCV station $< FCV_{mwt}$
EV minimum SoC after charging $> SoC_{min}^{EV}$
ICV minimum SoC after charging $> SoC_{min}^{ICV}$
FCV minimum SoC after charging $> SoC_{min}^{FCV}$
EV maximum number of simultaneous charging at EV station $< EV_{msc}$
ICV maximum number of simultaneous fueling at ICV station $< EV_{msf}$
FCV maximum number of simultaneous fueling at FCV station $< EV_{msf}$

8.5.3 Assumptions

Charging and fueling stations are reliable with fault-tolerant capabilities to work all
the time, with zero downtime. The presented model doesn't include the reliability
and protection models of the charging and fueling station.

8.6   Optimization Algorithm

Optimizing the hybrid charging station requires applying local optimization for each subsystem and implementing distributed optimization with negotiation between different views and parties. Figure 8.3 shows the proposed distributed optimization model where the optimization of each subsystem is performed based on local variables and objective functions while interfacing process variables with connected subsystems. The optimization of each subsystem is achieved by negotiation among the interconnected subsystems.

The optimization of the EV charging station is based on interactions with charging EV, MEG, grid, and TES, as shown in Fig. 8.4. Each EV can be charged from more than one EV charging station, based on local optimization within each EV, and each EV charging station. More than EV will be optimized while charged from the same EV station, with the consideration of incoming EVs. The interactions among

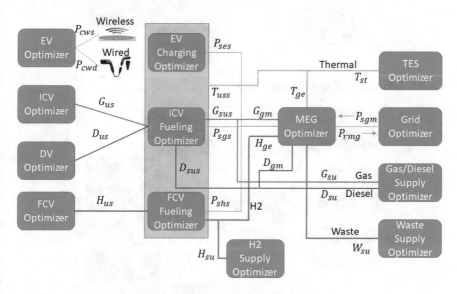

Fig. 8.3 Distributed optimization model for hybrid charging station

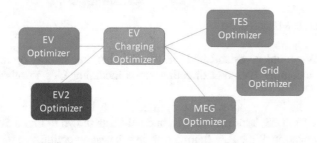

Fig. 8.4 Optimization of EV charging station

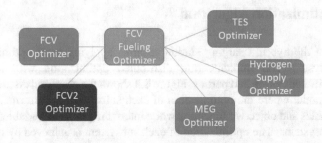

Fig. 8.5 Optimization of FCV charging station

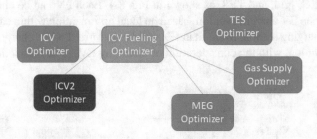

Fig. 8.6 Optimization of ICV charging station

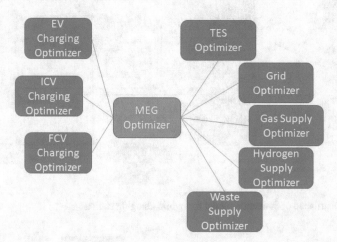

Fig. 8.7 MEG optimization model

incoming EVs could be established via the EV charging station, optimizing and prioritizing incoming EVs and directing some incoming EVs to other charging stations.

Similarly, the FCV fueling station optimization is based on optimization of all incoming FCVs, TES, hydrogen supply, and MEG, as shown in Fig. 8.5.

The optimization of the ICV fueling station is based on optimizing the incoming ICVs, TES, MEG, and gas supply, as shown in Fig. 8.6.

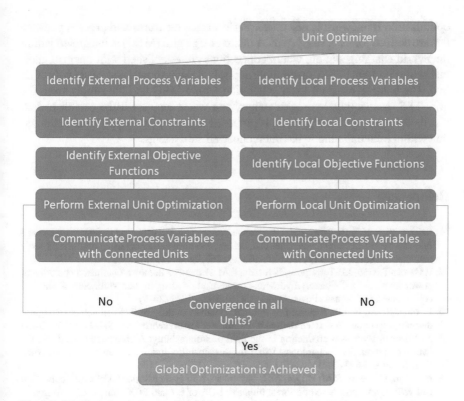

Fig. 8.8 Distributed optimization algorithm

The optimization of MEG is based on incoming energy supply from thermal, power, gas, and hydrogen, with the interactions of charging stations of EV, ICV, and FCV, as shown in Fig. 8.7.

The hybrid optimization framework is presented in Fig. 8.8, where optimization of each local unit is performed using local process variables, local constraints, and local objective functions while accepting process variables from all connected units. Similarly, each external unit will perform local optimization. Exchange of optimized process variables among connected units will be performed with iteration till convergence is achieved in all connected units.

8.7 Summary

This chapter discussed the possible design of a hybrid charging station that includes electric vehicle charging units, gas fueling units, diesel fueling units, and hydrogen fueling units for EVs, FCVs, DVs, and FCVs, respectively. The operation of the hybrid charging station is discussed. Related process variables are described as part of the design and operation. Optimization models are described, and an integrated optimization framework is discussed. Performance measures and process parameters are defined for each subsystem. A micro energy grid (MEG) is integrated within the hybrid charging station, where connections are established with the grid, thermal storage system (TES), gas/diesel supply, hydrogen supply, and waste collection line. The balance between local generation within MEG and the supply from the grid, TES, gas supply, hydrogen supply, and waste inputs will be established. A distributed optimization algorithm is proposed to ensure local and global optimization within each unit among the interconnected subsystems.

References

1. S. Ahmadi, S.M.T. Bathaee, A.H. Hosseinpour, Improving fuel economy and performance of a fuel-cell hybrid electric vehicle (fuel-cell, battery, and ultra-capacitor) using optimized energy management strategy. Energy Convers. Manag. **160**, 74–84 (2018)
2. P. García-Triviño, J.P. Torreglosa, F. Jurado, L.M. Fernández Ramírez, Optimised operation of power sources of a PV/battery/hydrogen-powered hybrid charging station for electric and fuel cell vehicles. IET Renew. Power Gener. **13**(16), 3022–3032 (2019)
3. P. Yang, X. Yu, G. Liu, Research on optimal operation strategy of fuel cell vehicle charging-discharging-storage integrated station. IOP Conf. Ser. Earth Environ. Sci. **675**(1), 12139 (2021)
4. N. Norouzi, Hydrogen production in the light of sustainability: A comparative study on the hydrogen production technologies using the sustainability index assessment method. Nucl. Eng. Technol. **54**(2) (2021)
5. B. Singh, G. Guest, R.M. Bright, A.H. Strømman, Life cycle assessment of electric and fuel cell vehicle transport based on forest biomass: LCA of EV and FCV using energy from biomass. J. Ind. Ecol. **18**(2), 176–186 (2014)
6. W. Yaïci, M. Longo, Assessment of renewable natural gas refueling stations for heavy-duty vehicles. J. Energy Resour. Technol. **144**(7), 1–12 (2022)
7. P. Song, Y. Sui, T. Shan, J. Hou, X. Wang, Assessment of hydrogen supply solutions for hydrogen fueling station: A Shanghai case study. Int. J. Hydrog. Energy **45**(58), 32884–32898 (2020)
8. X. Wu, H. Li, X. Wang, W. Zhao, Cooperative operation for wind turbines and hydrogen fueling stations with on-site hydrogen production. IEEE Trans. Sustain. Energy **11**(4), 2775–2789 (2020)
9. Z. Zsigraiová, G. Tavares, V. Semiao, Carvalho, M. de G, Integrated waste-to-energy conversion and waste transportation within island communities. Energy (Oxford) **34**(5), 623–635 (2009)
10. P. Hasanov, M.Y. Jaber, S. Zanoni, L.E. Zavanella, Closed-loop supply chain system with energy, transportation and waste disposal costs. Int. J. Sustain. Eng. **6**(4), 352–358 (2013)
11. A.R. Jordehi, Scheduling heat and power microgrids with storage systems, photovoltaic, wind, geothermal power units and solar heaters. J. Energy Storage **41**, 102996 (2021)
12. N. Danloup, V. Mirzabeiki, H. Allaoui, G. Goncalves, D. Julien, C. Mena, H. Allaoui, A. Choudhary, Reducing transportation greenhouse gas emissions with collaborative distribution: A case study. Manag. Res. Rev. **38**(10), 1049–1067 (2015)
13. EV Charging Data. https://evbox.com/en/ev-chargers/fast-charger. Accessed 13 Jan 2022
14. Canadian Fuels Association. https://www.canadianfuels.ca/our-industry/fuel-retailing/. Accessed 13 Jan 2022
15. Statistics Canada. Table 23-10-0066-01, Sales of fuel used for road motor vehicles, annual (x 1,000). https://doi.org/10.25318/2310006601-eng. Accessed 13 Jan 2022
16. U.S. Department of Energy. https://afdc.energy.gov/data/categories/fuel-consumption-and-efficiency. Accessed 13 Jan 2022
17. California Fuel Cell Partnership. https://cafcp.org/by_the_numbers. Accessed 13 Jan 2021

Chapter 9
Fast Charging for Marine Transportation

Hossam A. Gabbar

Acronyms

FCS   fast-charging station
GHG   greenhouse gas
MMR micro modular reactor
R&I   esearch and innovation

9.1 Introduction

Oceans and lakes around the world led to large waterfront areas linked to residential, industrial, recreational, mining, fishing, manufacturing, construction, and tourism areas. Social, commercial, and industrial activities are centered around waterfront areas. In North America, it is estimated that 30% of economic activities in United States/Canada are located in the Great Lakes region, with a population around 107 million, GDP US$ 6 trillion, and 51 million jobs [1]. Rivers and ocean regions are considered promising areas of research. This is especially true in energy production from the water current stream energy (WCSE) and wind energy, due to the availability of these resources throughout the year. There is an increasing demand for clean energy and smart and resilient infrastructures for both waterfront and maritime applications in Canada and worldwide. Energy deployment and planning in the maritime application are expanding due to the increase in world trade and shipping in view of worldwide development. Researchers have investigated

---

This chapter is contributed by Hossam A. Gabbar.

H. A. Gabbar (✉)
Energy Systems and Nuclear Science, University of Ontario Institute of Technology,
Oshawa, ON, Canada
e-mail: Hossam.gabbar@uoit.ca

© Springer Nature Switzerland AG 2022   147
H. A. Gabbar, *Fast Charging and Resilient Transportation Infrastructures in
Smart Cities*, https://doi.org/10.1007/978-3-031-09500-9_9

several economical ways to deploy energy systems in offshore areas and maritime applications [2]. Renewable energy systems, hydrokinetics, and hydropower technologies are potential maritime and waterfront applications. The analysis of current routes of maritime ships shows routes in different regions around the world, with heavy traffic for goods shipping and human mobility. Figure 9.1 shows marine traffic with a trip density as km/year, as described in [3].

The electrification of maritime supports ships in different types such as boats, yachts, cruises, naval vessels, supply vessels, offshore energy vessels, offshore platforms, oil and gas drilling rigs, icebreaking ships, and submarines. Ships' power system analysis includes modeling, simulation, control, protection, operation, communications, and optimization [4]. Optimization methods are investigated to ensure effective energy management for maritime applications [5]. Energy storage technologies are essential components in maritime applications' energy systems. They provide capabilities to store energy for long trips to be charged from different sources on board and during stops [6]. The charging of electric ships is established via wired or wireless charging technologies. Due to different ship and weather conditions, wireless charging of electric ships is promoted for a number of applications [7]. The wireless charging system is designed based on battery charging profiles and related parameters. Different wireless charging classes are defined in [8].

Fast charging is adopted in maritime applications. Tesla implemented fast-charging solution for maritime ships [9]. In Canada, the BCI Marine reported partnership with the Aqua superPower to install fast-charging points throughout Canada [10]. Fast charging will negatively affect voltage stability of power systems and the grid [11]. Design and operation considerations should be studied to reduce the impacts of fast charging on the power grid.

Maritime applications could be coupled with waterfront infrastructures. Energy networks are analyzed for waterfront applications, and examples from Toronto are presented in [12]. Renewable energy technologies are adopted widely in waterfront applications [13]. The analysis of energy systems for waterfront applications is

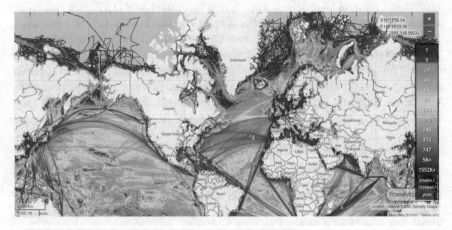

Fig. 9.1 Marine traffic profiles

performed based on sustainability [14]. The integration of renewable energy with diesel generators is favored in many applications, especially in areas that have access to fossil fuels. The integration with diesel generators is suitable during the energy transition where full coverage of energy requirements by renewables might not be achieved using only renewable due to high costs. An example of the optimized design of a hybrid PV-wind-diesel energy system in Bangladesh is presented for coastal areas [15].

The development of hybrid energy systems for maritime and waterfront applications will reduce deployment costs of charging stations for maritime applications. In addition, it will support the expansion of hybrid energy systems with more penetration of renewable energy technologies in maritime and waterfront applications. The following section will explain the functional modeling of a hybrid energy system integrated with charging infrastructures, which will support both electrifications of maritime ships and sustainable development of waterfront regions.

9.2 Functional Modeling of Hybrid Energy System for Maritime and Waterfront Applications

Maritime electrification is crucial to reducing GHG emissions from marine transportation. The planning of novel and practical charging solutions requires a proper understanding of functions involved in the charging infrastructure. Figure 9.2 describes the detailed functional modeling of an integrated solution for an energy system with charging infrastructure for maritime and waterfront applications. The hybrid energy system is defined based on grid interface, renewable energy systems, hydrokinetic system, hybrid energy storage, and power electronics to integrate

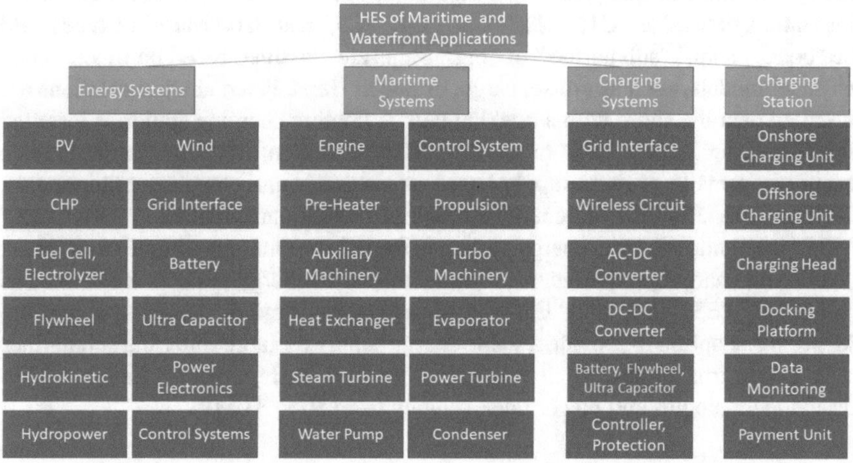

Fig. 9.2 Functional modeling of hybrid energy system for maritime and waterfront applications

hybrid energy system with grid, ship, and charging station. The main maritime functions related to energy and charging are listed, including engine, turbine, heat exchanger, evaporator, water pump, auxiliary machinery, and control system. The main functions of the charging station include onshore unit, offshore unit, charging head, data monitoring, and payment unit. The charging system includes a grid interface, charging head, AC-DC converter, DC-DC converter, wireless circuit, energy storage of battery/ultracapacitor/flywheel, and controllers. Functions related to waterfront applications include the interface of a hybrid energy system with facilities such as houses, hotels, cottages, parks, recreational buildings, airports, seaports, factories, nuclear power plants, or fish farms. The charging station in the waterfront areas will support the peak shift of energy supply and charging EVs and boats. Excess energy from the hybrid energy system is used to supply energy to loads and communities in the waterfront areas.

9.3   Energy System Design for Maritime and Waterfront

The energy system design is defined to support the charging station, the maritime ship, and the waterfront applications. The energy system design is based on integrating grid, renewable energy system, and energy storage.

9.3.1   Energy System Design Scenarios

The configuration of the hybrid energy system for the charging is defined based on maritime charging load profile and load profile from the waterfront area, as shown in Fig. 9.3. There are six possible configurations for the energy system for the charging station, defined as "C1," "C2," "C3," "C4," "C5," and "C6." Maritime type could be "boat," "ship," "submarine," or "tug." Ships are classified based on function and physical models, such as cruise, cargo, or tanker ships. Fixed charging stations are installed near the shore with a docking unit. A floating charging station is installed in offshore areas as a static or moving offshore station. Maritime might include energy systems to provide supply based on shiploads and charging requirements. Different scenarios for the energy storage in the ship are defined as "S1," "S2," "S3," "S4," and "S5." The energy system in the ship is configured based on shipload profile and charging requirements. Configurations of energy systems are defined as "E1," "E2," "E3," "E4," and "E5." A nuclear-renewable hybrid energy system using MMR is one option to provide a stable energy supply for large ships and submarines with longer trip ranges. Nuclear reactors are used in ships with different reactor design technologies and power rates, ranging from 50 to 35 MW.

9.3.2 Performance Measures

The evaluation of different design configurations is performed using performance measures, as explained in Table 9.1.

Specific parameters of the ship are defined in Table 9.2. These parameters are defined for each ship based on trip details (Table 9.3).

Nuclear reactor parameters are defined in Table 9.4.

Parameters for the PV are defined in Table 9.5.

Wind turbine parameters are defined in Table 9.6.

Parameters for the battery energy storage are defined in Table 9.7.

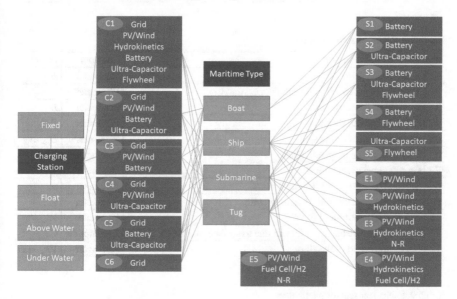

Fig. 9.3 Hybrid energy system design scenarios for maritime and waterfront application

Table 9.1 Performance measures

Cost of energy (COE)
Internal rate of return (IRR)
Net present value (NPV)
Loss of power supply probability (LPSP)
Surplus energy fraction (SEF)
Level of autonomy (LA)
Levelized cost of energy (LCOE)
Lifecycle cost (LCC)
Total annualized cost (TAC)
Loss of load probability (LPP)
Generation reliability factor (GRF)
CO_2 gas emissions
Power-to-weight ratio (PWR)

Table 9.2 Ship model parameters

Ship's name (IMO number)
Date delivered/builder (where built)
Flag/port of registry
Type of ship
Length overall (LOA)
Length between perpendiculars (LBP)
Extreme breadth (beam)
Deadweight
Displacement

Table 9.3 Detailed ship parameters

#	Parameter/assumption	Category	Notation
1	Beam of the ship	Parameter	B
2	Volume displacement of the ship	Parameter	v
3	Draught of the ship	Parameter	D
4	Extreme breadth (beam)	Parameter	Bex
5	Average draught of the ship	Parameter	D_avg
6	Length between perpendiculars	Parameter	LBP
7	Gravitational acceleration	Parameter	g
8	Seawater density at 30 °C temperature	Parameter	ρ_w
9	Seawater viscosity at 30 °C temperature	Parameter	γ_w
10	Average speed of the ship	Parameter	Vs_avg
11	Incremental resistance coefficient due to surface roughness of ship	Assumption	C_A
12	Maximum speed of the ship	Parameter	Vs_max

Table 9.4 Nuclear reactor parameters

Parameters
Reactor size (kWe)
Plant lifetime (years)
Overnight capital cost ($/kWe)
Fixed operation and maintenance (O&M) cost ($/kWe)
Fuel cost ($/MWh)
Refueling cost of fuel module ($)
Core lifetime (years)
Decommissioning cost ($/MWh)
Capacity factor (%)
Plant efficiency (%)
CO_2 emission (kg/MWh)

Table 9.5 PV parameters

Parameter
Area occupied by unit PV panel (m^2)
Capital cost ($/kW)
Replacement cost ($/kW)
O&M cost ($/kW/year)
Lifetime (years)
Reference efficiency of PV panel (%)
Efficiency of the MPPT unit (%)
Temperature coefficient ($°C^{-1}$)
PV panel reference temperature (°C)
Nominal operating cell temperature (°C)

Table 9.6 Wind turbine parameters

Parameter
Nominal capacity (kW)
Capital cost ($/kW)
Replacement cost ($/kW)
O&M cost ($/kW/year)
Lifetime (years)
Hub height (m)
Anemometer height (m)
Minimum wind speed (m/s)
Maximum wind speed (m/s)
Rated wind speed (m/s)
Power law exponent

Table 9.7 Battery storage parameters

Parameter
Capital cost ($/kWh)
Replacement cost ($/kWh)
O&M cost ($/kWh/year)
Lifetime (years)
Efficiency (%)
Days of autonomy (days)
Depth of discharge (%)
Inverter efficiency (%)

Fig. 9.4 Ship route with maritime charging stations

9.3.3   Ship Route

The location and type of maritime charging stations are planned and optimized based on trip routes. Figure 9.4 shows one trip route of "BALTIC SUNRISE," which has parameters defined as maximum speed of 17.8 km with average speed of 11.94 km. The maximum draught and minimum draught of the ship were 21.6 m and 10.7 m, respectively [16].

9.3.4   System Design

The proposed hybrid energy system with possible configuration options of energy and storage systems includes renewable energy systems, nuclear reactors, energy storage systems, and power electronics. The bidirectional DC-DC converter supports the charging of maritime such as boats, ships, submarines, and tugs. AC and DC loads from waterfront applications are linked to the energy system. Effective energy management will balance loads, storage, and generation in addition to the link to the grid. Figure 9.5 shows the integrated design covering different configurations and load profiles from maritime and waterfront applications.

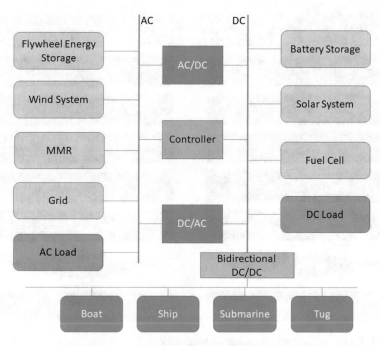

Fig. 9.5 Design of charging station for maritime and waterfront application

9.3.5   Optimization

The optimization of design and operation is achieved for the integrated hybrid energy system, charging station, charging loads, and AC/DC loads of different energy system design configurations for charging stations and maritime ships. The optimization is performed during the system design stage and operation for every incoming ship charging request or waterfront load profile. The detailed optimization framework is shown in Fig. 9.6. Parameters are defined for each physical system, including battery, ultracapacitor, flywheel, PV, wind, nuclear reactor, power electronics, grid interface, charging system, maritime system components, and maritime energy storage system. The optimization is based on predefined objective functions that translate all performance measures into net profit, which shows the difference between expenses and income. The optimization is achieved by maximizing net profit for each system: ship, charging station, and utility (grid). Income includes charging costs and sending energy back to the grid. Expenses include the cost of energy, maintenance cost, operating cost, and losses. Net profit is defined for each view and optimized locally while coordinating global optimization with weight factors. Constraints are defined for each component which are applied to the optimization process. Constraints are defined for grid supply, PV, wind, nuclear reactor, and hydrokinetics. Constraints are defined for battery performance, including SoC (state of charge). The ship's carrying capacity limit is defined by the weight of energy

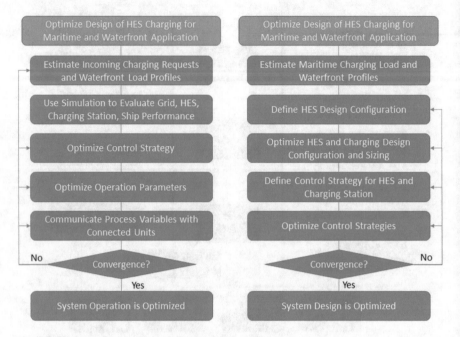

Fig. 9.6 Design and operation optimization framework

system, storage, and cargo weight and included in the constraints and linked to speed and energy requirement. Area limitation to installing PV and wind in the station side is limited with space constraints. Similarly, the area in the ship for PV installation is limited by the space available in the ship.

Optimization decision variables are defined as per Table 9.8.

9.3.6   Cargo and Propulsion Modules for Nuclear-Powered Ships

Nuclear power in ships and maritime applications covered limitations of long-range trips and GHG emissions from gas-based ships. However, some countries have restrictions to accept ships powered with nuclear reactors due to national policies and regulations. In these countries, the entrance of ships to their local territories and seaports is restricted. To address this challenge, a modular design of a nuclear-powered ship is used with two main modules: propulsion module and cargo module. The propulsion module can be separated from the ship before entering the national territory. The propulsion module contains the nuclear plant, whereas the cargo module carries the payload. The propulsion module will remain in the international water till the cargo module completes planned activities in the national territory and seaport. A secondary propulsion system within the cargo module can be used to move the ship within the national territory. The secondary propulsion can include

Table 9.8 Optimization decision variables

Grid	Energy supplied
	Energy returned
Charging station	PV size
	Wind size
	Hydrokinetics
	Battery type and size
	Ultracapacitor type and size
	Flywheel type and size
	MMR model and size
Ship	PV size
	Fuel cell type and size
	Hydrokinetics
	Battery type and size
	Ultracapacitor type and size
	Flywheel type and size
	MMR model and size

renewable energy systems placed locally in the cargo module or a tug unit. Conventional- or fossil fuel-based energy systems can be used in some cases to move the cargo module. Figure 9.7 shows the separate modules with possible separation to enter the restricted national territories.

9.4 Advances in Research and Innovation

Research and innovations are among the key success factors to technology development and practical deployments and implementations. The advancement of marine transportation electrification and fast-charging infrastructures requires novel ideas and lab testing to transform advanced lab facilities to support different research activities. There are several research facilities dedicated to supporting marine engineering and safety. Other research facilities are dedicated to supporting power systems for marine, such as motors, drives, and engines. However, a limited number of research facilities are dedicated to supporting innovations in energy systems for the marine sector. An integrated research facility will support hybrid energy systems and charging for marine and waterfront applications.

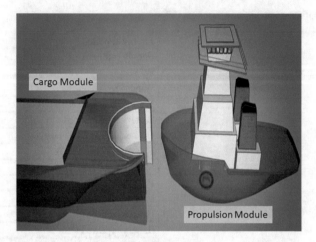

Fig. 9.7 Modular cargo and propulsion for maritime ships

9.4.1   Research on Energy Systems for Marine Transportation and Waterfront Infrastructures

Many researchers worldwide are investigating advanced research and innovations to promote energy systems for maritime and waterfront applications. Floating PV, hydrokinetics, hydropower, and wind systems are the main components in energy systems. The integration among different technologies within the hybrid energy system will support waterfront regions. The research should support advances in maritime applications, such as autonomous boats, the safety of boats, and electric motors. The research facility can include water flume for testing underwater energy systems, charging, stability during charging, and safety under harsh and severe weather conditions. Figure 9.8 shows the main subsystems for integrated research and test facility.

9.4.2   Research Areas

Research and innovation cover main areas, as shown in Fig. 9.9. It includes onshore and offshore micro energy grids, grid integration, hybrid energy systems, efficiency, electric motors, drives, machines, waterfront planning, collection and treatment of waste from oceans and lakes, charging infrastructures, wireless charging for boats, and safety assessment of energy systems and charging stations. The research can include nuclear-renewable hybrid energy system integration and applications. The research supports innovations and technology development for several applications, including construction, fishing, mining, transportation, wind turbine technology, hydrokinetic technology, dam structures, and ship development.

Fig. 9.8 R&I of hybrid energy system for marine and waterfront applications

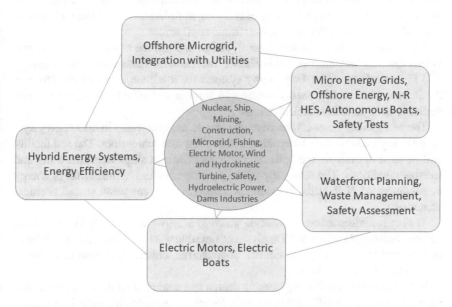

Fig. 9.9 Research and innovation for maritime and waterfront applications

The research and test facility will support research activities and promote industries to develop and test technologies and is also used to train professionals and trainees. The research facility supports water flow analysis, energy systems, and the coupling between water and energy systems. It also supports maritime transportation and charging systems. The analysis of charging and fast charging in different water, weather, and ship conditions supports design, control, and technology development. The testing includes monitoring, testing, and data analytics. The facility supports key parameters related to water, energy, and transportation. The research facility includes co-simulation to enable real-time testing of parameters from the research facility and evaluate different charging parameters under different conditions. The integrated co-simulation includes water flow simulation and the energy system simulator with HOMER and MATLAB and real-time simulation with OPAL-RT. The proposed simulator allows water flow with different speeds linked to wind speed and weather conditions to test energy systems and charging mechanisms, as well as waterfront infrastructures. The facility tests energy system innovations such as hybrid wind turbines and hydrokinetic systems. The control and simulation models are developed based on different water current waves and wind energies. Electric motors are tested using experiments and simulation practices in different energy supply profiles and water conditions.

The research facility uses the simulator to design and test electric motors and their integration within electric boats, ships, and submarines in different water and weather conditions, including different currents, waves, and wind speeds. The programmable current and wave profiles allow assessing the design and efficiency of electric motors and optimize their performance in different operation scenarios. It also supports the design and testing of charging stations based on different load and demand profiles for waterfront and maritime applications.

The research utilizes the simulator to design and test the novel wind and hydrokinetic turbines for waterfront and offshore installations. It integrates a micro energy grid for waterfront and hybrid energy systems for maritime applications. Floating PV is integrated within energy systems to support waterfront infrastructures and maritime transportation electrifications.

The test facility supports the planning of nuclear-renewable hybrid energy systems (N-R HES) and their integration within ships and submarines. The test facility will be used to evaluate the optimal position of SMR in ships based on water and weather conditions. The tests support the development and optimization of energy management of N-R HES. It evaluates safety in different weather and accident scenarios.

Testing is performed on small-scale waste-to-energy units suitable for waterfront and maritime applications with analyses of possible waste in lakes and the offshore regions and the best ways to collect, characterize, collect, and treat in different water and weather conditions.

It is used to support the engineering design and testing of autonomous boats and maritime systems in different conditions for many applications, including rescue, recreation, manufacturing, construction, mining, health, agriculture, fishing, and shipping. It can be used to test integrated drones with autonomous boats for waterfront and maritime applications.

It is used to test waterfront infrastructures in different weather and water conditions, including flood scenarios; evaluate safety and health measures; and support training for municipalities and the public. It supports the design and testing of propellers for ships and yachts and the testing of the interactions between yachts and ship berths.

The interactions among the research groups will support the research and innovations with enhanced technology development and testing. The main research groups include energy system, mechanical, electrical, nuclear, transportation, aerospace, sustainability, and data analytics.

9.4.3   Research and Test Facility

The research facility includes the following main components: waterfront infrastructure simulator to reflect load profiles, generation profiles, weather profiles, and waste generation profiles and link to energy systems from offshore and waterfront systems. It includes e-motor test bench, programmable automatic load and source units, programmable inverter cards, power electronics, power converters, serial communication modules/cards for communication protocols, lithium-ion battery bank with programmable battery management card, battery management system (BMS), and interface systems, wind turbine, hydrokinetic turbine, hydroelectric power motor, and power electronic components including transducers, sensor, current transformers (CT), and potential transformers (PT). A water flow simulator is used to provide input profile for microgrid and hybrid energy system design and optimization. High-speed fan is used to emulate the wind energy. The simulation workstation is used to simulate different flow rates of pumping water to match water velocity. The simulation workstation should include a wave generator modeler and evaluation of other parameters such as water speed, wave height, flow rate, power generated, and hydraulic power components, as well as design parameters of energy systems, as shown in Fig. 9.10. A wind turbine is employed to test different energy generation profiles. The hydroelectric power engine/motor will be tested based on

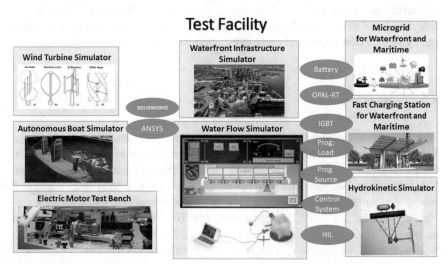

Fig. 9.10 Research and test facility

different profiles. Data will be fed back to the integrated setup with hardware-in-the-loop (HIL) simulation modules, which has monitoring, control, and co-simulation modules.

It includes an integrated simulator with the following modules: computational fluid dynamics (CFD), real-time power simulation using OPAL-RT, and engineering design using SOLIDWORKS. The proposed facility will have the first integrated hybrid energy system for waterfront and maritime applications.

9.4.4   Research Impacts

The research facility supports the development of advanced hybrid energy systems and fast-charging unit for waterfront and maritime applications. This has high impacts on reduced GHG emissions and reduced dependency on the grid. It leads to reduced electricity use and costs. It contributes to economic development for offshore and waterfront regions. Other benefits include the increase of renewable energy penetration. It also enhances air and water quality and reduces environmental stresses. The energy and transportation research for maritime and waterfront applications will enhance economy and community development.

9.4.5   Target Industries

The research will target the following industries: ship builder, ship operator, maritime organization (national and international), energy and nuclear regulatory organizations, energy and nuclear safety organizations, construction companies, mining companies, offshore petrochemical companies, motor manufacturing companies, energy system developers and integrators, energy utilities, and municipalities.

9.5   Summary

This chapter discussed fast-charging infrastructures for maritime applications. Renewable energy systems are integrated within maritime systems and charging infrastructures to support transportation electrification in maritime applications. The integration between charging infrastructures for maritime and waterfront energy systems is presented using a number of possible design and operation scenarios. Sustainability considerations and reduced GHG emissions are discussed and included in the design and operation of charging infrastructures for maritime applications. Functional modeling is presented for hybrid energy systems for maritime transportation electrification as integrated with waterfront applications. Renewable and energy storage technologies are integrated to meet load profiles for maritime charging and waterfront energy supply demands. Model parameters are identified for the integrated charging station configurations with maritime ships to meet demand shipping routes and charging load profiles for different routes. An integrated framework and techniques are discussed to optimize design and operation functions. Performance measures are defined for each component of the integrated system. Research and innovation best practices are illustrated to support the development of fast-charging infrastructures for maritime and waterfront applications.

References

1. The Great Lakes Economy: The Growth Engine of North America, Online, https://councilgreatlakesregion.org/the-great-lakes-economy-the-growth-engine-of-north-america/. Accessed 27 Jan 2022
2. L. Wang, C. Liang, J. Shi, A. Molavi, G. Lim, Y. Zhang, A bilevel hybrid economic approach for optimal deployment of onshore power supply in maritime ports. Appl. Energy **292**, 116892 (2021)
3. Marine Traffic, Online, https://www.marinetraffic.com/en/ais/home/centerx:-12.0/centery:24.8/zoom:2. Accessed 27 Jan 2022
4. G. Sulligoi, A.K. Rathore, Guest editorial marine systems electrification. IEEE Trans. Aerosp. Electron. **2**(4), 504–506 (2016)
5. R. Tang, X. Li, J. Lai, A novel optimal energy-management strategy for a maritime hybrid energy system based on large-scale global optimization. Appl. Energy **228**, 254–264 (2018)
6. Z. Zhou, M. Benbouzid, J. Frédéric Charpentier, F. Scuiller, T. Tang, A review of energy storage technologies for marine current energy systems. Renew. Sust. Energ. Rev. **18**, 390–400 (2013)
7. C. Yajie, P. Zhiqiang, N. Ming, G. Haibo, P. Zhao, H. He, Z. Yihang, Design of marine high power wireless charging system. E3S Web Conf. **194**, 02010 (2020)
8. M. Liu, C. Zhao, J. Song, C. Ma, Battery charging profile-based parameter design of a 6.78-MHz class E^2 wireless charging system. IEEE Trans. Ind. Electron. (1982) **64**(8), 6169–6178 (2017)
9. B. Bundale, *Tesla to Put Fast-Charging Stations Throughout Maritimes, Chronicle-Herald* (Halifax, N.S., 2018)
10. BCI Marine Partners with Aqua Superpower to Install Fast-Charging Points Throughout Canada, Canada NewsWire (2022)
11. C.H. Dharmakeerthi, N. Mithulananthan, T.K. Saha, Impact of electric vehicle fast charging on power system voltage stability. Int. J. Electr. Power Energy Syst. **57**, 241–249 (2014)
12. S. Prudham, G. Gad, R. Anderson, J. Laidley, G. Desfor, Networks of power: Toronto's waterfront energy systems from 1840 to 1970, in *Reshaping Toronto's Waterfront*, (2017), pp. 175–200
13. Renewable Energy Systems: Winning Place on the Waterfront, What's New in Building, p. 46 (2008)
14. S.S. Eldeeb, A.G. Rania, A.E. Sarhan, A sustainability assessment framework for waterfront communities. Renew. Energy Sustain. Dev. **1**(1), 167–183 (2015)
15. S. Rashid, S. Rana, S.K.A. Shezan, A.B.K. Sayuti, S. Anower, Optimized design of a hybrid PV-wind-diesel energy system for sustainable development at coastal areas in Bangladesh. Environ. Prog. Sustain. Energy **36**(1), 297–304 (2017)
16. Auke Visser's International Super Tankers, http://www.aukevisser.nl/supertankers/VLCC %20T-V/id1147.htm. Accessed 17 Oct 2020

Chapter 10
Resilient Charging Stations for Harsh Environment and Emergencies

Hossam A. Gabbar

Acronyms

BBN   Bayesian belief network
CU   charging unit
FCS   fast-charging station
SoC   state of charge
V2G   vehicle to grid
V2H   vehicle to home

Nomenclature

EV_{msf} Maximum number of simultaneous FCV fueling at FCV station

10.1 Introduction

The world is witnessing climate change, leading to several extreme weather conditions such as volcanic eruptions, snow storms, very low temperatures, wind storms, freezing rain, tornados, tsunami, and harsh ocean waves. These extreme weather conditions and harsh environments led to many social, health, economic, and environmental consequences. Harsh environments are negatively impacting electronic components and systems [1]. Electronics are included in many applications such as

This chapter is contributed by Hossam A. Gabbar.

H. A. Gabbar (✉)
Energy Systems and Nuclear Science, University of Ontario Institute of Technology,
Oshawa, ON, Canada
e-mail: Hossam.gabbar@uoit.ca

© Springer Nature Switzerland AG 2022
H. A. Gabbar, *Fast Charging and Resilient Transportation Infrastructures in
Smart Cities*, https://doi.org/10.1007/978-3-031-09500-9_10

power and energy, industry, transportation, aerospace, maritime, and health. Any system that operates outdoor is subject to a harsh environment. Charging stations are important infrastructures that operate outdoors and can be seriously impacted in a harsh environment. Most of the designs of fast-charging stations include PV. Harsh environment could be translated into very high or low temperature, high wind speed, or snow storm. These have negative impacts on power generated from PV, which will lead to reduced energy capacity in the charging station [2]. From the other side, severe conditions might lead to higher charging demands due to possible interruptions in transit systems. The reduced energy capacity in charging stations and increased charging demand profiles will lead to higher risks to the transportation network and mobility requirements. Fires are other severe conditions that might negatively impact power lines that support charging infrastructures. Fires are caused by events related to the triangle of fire: oxygen, heat, and fuel. Batteries might overheat due to a number of internal or external reasons [3]. Once the battery becomes overheated, this might lead to fire and other health concerns, in addition to interruptions to charging stations. The uncertainty of these events could be quantified using probability calculation. Bayesian belief network (BBN) is used to estimate the probability of dependent events. BBN is applied on the integrity and reliability of physical systems of power and energy systems, such as estimating the probabilistic health index of power transformer [4]. Power quality and harmonics could be estimated using probabilistic calculation, part of distribution networks. EV charging stations are integrated with power distribution networks, and power quality impact estimation is evaluated using BBN [5]. Other techniques such as Monte Carlo simulation are applied on wind-powered EV charging stations to estimate the probability of operation modes and security analysis [6]. Probability of excess voltage in a distribution network is estimated to support balanced operation of charging EV load [7]. Resiliency assessments have been conducted on transportation electrification. Example of techniques applied on electrified road networks subject to charging station failures [8]. Traffic demands are analyzed prior to inputs to the charging station. Different failures of the charging station could be analyzed, where distributed or decentralized fault tolerance strategy could be defined and optimized for different subsystems of the charging station [9]. AI techniques such as particle swarm optimization (PSO) are used to optimize the charging station's safe operation [10]. Failure of power system components could lead to interruptions to the charging station. Integrity assessment and reliability analysis of physical components such as the transformer will indicate early scheduling and managing incoming EV charging requests. The two-way communications between utility substations and EV charging stations will balance utility substation conditions based on incoming charging requests [11]. Actions from the decision process could include the rescheduling of charging requests. The decision of the target SOC could be evaluated as well. Also, maintenance and inspection activities could be scheduled based on the condition of a physical component such as a transformer and considering charging load profiles. Computation and graph methods are used to predict the outage of the utility grid based on ultrafast-charging loads [12]. Safety considerations are analyzed for large-scale charging stations [13]. Potential safety considerations include facility degradation, cyberattacks on the charging station, and unbalance between the energy capacity in the charging station using renewable energy and the charging loads. These disturbances will lead to instability in normal operation. Some of the incoming charging loads could be balanced with V2G in selected stations, such as idle stations [14]. The resilience analysis framework is defined as layer of resilience analysis (LORA), which includes inherent resiliency design, resiliency control, resiliency alarms, and resiliency interlocks [15]. Resiliency functions could be implemented using function added to a component, such as bidirectional converter, to enable specified functions such as V2G or V2H with capabilities such as emergency discharging functions [16]. Researchers investigated emergencies and analyzed the causation scenarios to provide suitable procedures for charging stations [17]. The recovery process could be designed based on resiliency scenarios. The mapping recovery process with resiliency is applied on power distribution networks while considering transportation charging loads [18].

10.2 Charging Infrastructures

Charging infrastructures can be flexibly designed based on transportation and mobility demand. Figure 10.1 shows flexible models for charging infrastructures in three main models. The first model is charging units (CU) at any location, such as a building, house, mall, or park. The second model is charging while moving using wireless V2V or V2I. The third model is a charging station for both EV and battery (BT), where charging is performed with wireless or wired charging. Battery swap and charging can also be performed in this model. The proposed flexible charging model will allow alternative models such as swap battery, charge at different locations, or share EV.

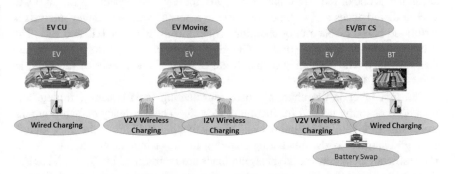

Fig. 10.1 Flexible models for charging infrastructures

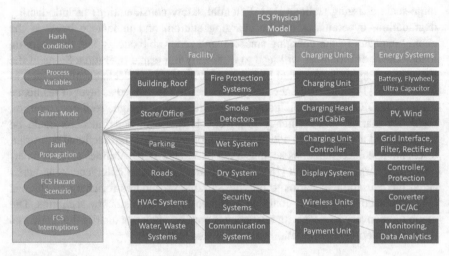

Fig. 10.2 Modeling harsh environment for fast-charging station

10.3   Charging in Harsh Environment

A harsh environment is modeled with a set of process variables such as temperature, smoke concentration, and wind speed (Fig. 10.2). The different process variables are linked to the failure mode of components within FCS. The possible fault propagation scenarios are defined based on possible hazard scenarios of the FCS. The evaluation of possible FCS interruptions is quantified based on the probabilities of different events. Bayesian belief network (BBN) is used to estimate probabilities.

10.4   Resiliency Analysis of Charging Infrastructures

Resiliency is one of the performance measures for any given system as the ability to react to disturbances and resume operation in a normal way. Resiliency analysis is performed based on incoming charging requests from EV, EB, ET, and EM. The charging requests are for wireless charging, wired charging, swap battery, and charge battery. The resiliency demands are defined at each layer of the FCS, as shown in Fig. 10.3.

The first layer is the charging load, representing the frequency, timing, and amount of charging needed for incoming requests. The resiliency demands are the probability of incoming imitating events for activating resiliency functions, including high charging loads, fluctuating charging loads, or increasing charging loads. The resiliency demands for the charging loads are represented by "RD11," RD12," "RD13," and "RD13." The resiliency demands for the charging unit are defined as

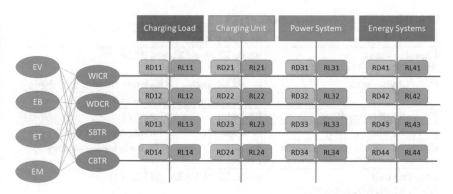

Fig. 10.3 Resiliency analysis of FCS

"RD21," "RD22," "RD23," and "RD24." The resiliency demands for the power system are defined as "RD31," "RD32," "RD33," and "RD34." Similarly, the resiliency demands for the energy systems are defined as "RD41," "RD42," "RD43," and "RD44." Resiliency layers are defined for each resiliency demand to resolve or mitigate the incoming stresses based on the corresponding resiliency demand. The resiliency layer is defined with a probability of success to improve resiliency for incoming resiliency demands. Target resiliency is the system's performance in view of incoming resiliency demands and the successful application of resiliency layers. Resiliency analysis examples are shown in Table 10.1.

The protection and safety functions are designed to overcome hazard conditions that will cause damage to charging station subsystems. Some of the root causes of the protection systems could be resolved via resiliency layers, which might mitigate the root causes of the initiating event of the demand to the protection layer.

Resiliency layer analysis is shown in Fig. 10.4. Resiliency demands are and layers are mapped to FCS. Resiliency demands for the incoming charging requests are defined as "RD1*." Resiliency demands for charging units are defined as "RD2*" and resiliency layers are defined as "RL1*." Resiliency demands for the power systems are defined as "RD3*" and resiliency layers are defined as "RL3*." And resiliency demands for the energy systems are defined as "RD4*" and the resiliency layers are defined as "RL4*." Layer of resiliency analysis or "LORA" is a framework to support the design of resiliency layers in view of resiliency demands. A probabilistic approach is used to assess each resiliency layer in view of the probability of resiliency demands and target resiliency with resiliency reduction factors. For example, if resiliency demand is evaluated as RD1, target resiliency is TR1. Then RL1 should be selected to cover the gap between RD1 and RL1.

Resiliency analysis will include the following main concepts: inherent resiliency design (IRD), resiliency control system (RCS), resiliency alarm management (RAM), and resiliency interlock systems (RIS). Some resiliency functions could be implemented as inherence resiliency features, such as the layout of the station to meet high-income charging vehicles. Resiliency functions could be implemented as part of the control system, such as limiting power to meet grid conditions or

Table 10.1 Resiliency analysis examples of FCS

Resiliency demand	Resiliency layer
Charging loads **RD11, RD12, RD14,** and **RD14**	**RL11, RL12, RL14,** and **RL14**
High charging loads	Revise scheduling for improved charging load shifting
	Increase charging capacity of FCS
Fluctuating charging loads	Revise scheduling for improved charging load shifting
Increasing charging loads	Redesign or increase charging capacity of FCS
Congested roads to charging station	Revise traffic planning for roads leading to charging station
Charging unit **RD21, RD22, RD23,** and **RD24**	**RL21, RL22, RL23,** and **RL24**
Low visibility at charging unit	Use automated charging unit
	Shift to wireless charging
Strong wind at charging unit	Increase testing, inspection, and physical support of the charging unit
	Revise station layout to reduce wind impact on station
Power system **RD31, RD32, RD33,** and **RD34**	**RL21, RL22, RL23,** and **RL24**
High temperature of power electronic systems	Design of thermal management of power electronic systems
	Balance load on power electronics by shifting charge load
Energy system **RD41, RD42, RD43,** and **RD44**	**RL41, RL42, RL43,** and **RL44**
High charging load	Increase energy capacity and storage of FCS
Frequent failure of the controller	Design fault-tolerant control

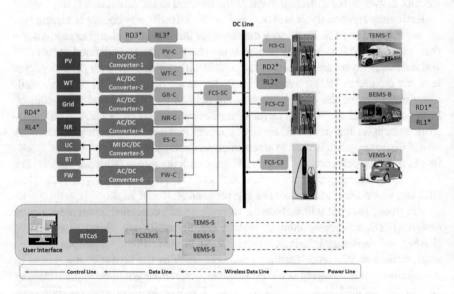

Fig. 10.4 Resiliency layer analysis of FCS

charging demand. Resiliency functions could be implemented in the form of alarms to send automated alerts about storms, high demand, or grid interruptions. Resiliency functions could be implemented as interlock systems, such as an automatic current limiter per battery to allow balanced charging at different charging units.

Resiliency analysis is performed using independent resilience layers (IRLs), which are used as part of the layer of resilience analysis (LORA) framework. Resiliency analysis is explained in Fig. 10.5. Resiliency demands are defined for different sections, charging loads, charging units, power systems, and energy systems. The resiliency demand is defined as "RD1*," which is a probability or frequency of event per year. The resiliency layer is used to achieve resiliency in view of resiliency demand and defined as "RL1*." The second resiliency demand is "RD2*," which is mitigated with resiliency layer "RL2*." Resiliency layer "RL2*" can also support the mitigation of resiliency demand "RD1*."

Similarly, resiliency layers are defined for "RL3*" and "RL4*." An example is used with values to show each resiliency demand and layer possible probabilities. The target resiliency is the desired performance in view of input resiliency demands. If the current resiliency level is below the target resiliency, it will be required to improve one or more of the presented resiliency layers with a high success probability. Final mitigated resiliency is defined as "TR*."

For example, the probability of resiliency demand "RD1*" is 0.05. The success factor of resiliency layer "RL1*" is 0.97. These values are evaluated based on monitoring data and component reliability. Similarly, values are defined for "RD2*," "RD3*," "RD4*," "RL2*," "RL3*," and "RL4*." The multiplication of "RD1*" by "RL1*" will lead to the probability of a successful resiliency layer in view of incoming resiliency demand, which is of the value of 0.0485. The second resiliency layer is evaluated as ("RD1*" x "RL1*") + ("RD2*"), which will be multiplied by "RL2*" to produce the resiliency mitigation after "RL2*." Similarly, other resiliency layers are evaluated, and the total equation is presented in Eq. (10.1):

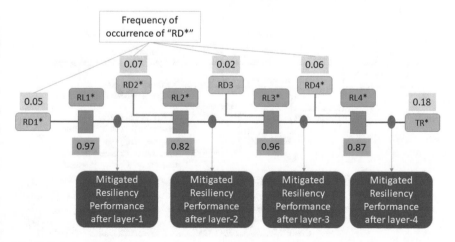

Fig. 10.5 Independent resilience layers (IRLs) with resiliency demand and resiliency layers

$$\text{Final mitigated resiliency} = \left(\left(\left(\left(\left(\text{"RD1*"} \times \text{"RL1*"}\right) + \text{"RD2*"}\right) \times \text{"RL2*"}\right) + \text{"RD3*"}\right) \times \text{"RL3*"}\right) + \text{"RD4*"}\right) \times \text{"RL4*"} \quad (10.1)$$

Based on the presented example, the final mitigated resiliency after "RL4*" = $((((((0.05 \times 0.97) + 0.07) \times 0.82) + 0.02") \times 0.96) + 0.06) \times 0.87 = 0.15$, which is reflected into the final resiliency performance. The target mitigated resiliency is defined in this example as 0.18, which is higher than current resultant resiliency performance of 0.15. Hence, the resiliency design should be revised to improve the system resiliency by changing resiliency layers to achieve higher performance. For example, if we enhance all resiliency layers to be 0.98 (i.e., "RL1*" = "RL2*" "RL3*" = "RL4*" = 0.98), the final mitigated resiliency after "RL4*" will be = $((((((0.05 \times 0.98) + 0.07) \times 0.98) + 0.02") \times 0.98) + 0.06) \times 0.98 = 0.19$, which is higher than the target resiliency performance of 0.18. Hence, the proposed enhanced resiliency will lead to an acceptable resiliency performance.

In order to understand the detailed layer of resilience analysis (LORA) of the target station, an example is illustrated as shown in Fig. 10.6.

The layer of resilience analysis will enable designers to evaluate incoming resilience demands at each layer of the station, including charging load, charging unit, power system, and energy system. The frequency of each resiliency demand is quantified, and the corresponding probability of failure of each resiliency later is defined as PFD1, PFD2, etc. The successful mitigation of resiliency demand is represented as (1 – PFD). The unsuccessful mitigation of resiliency demand will lead to unacceptable resiliency performance. It might also lead to a protection demand if

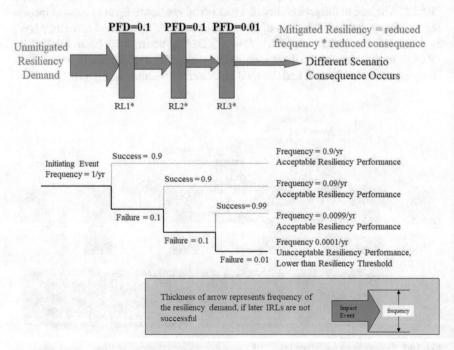

Fig. 10.6 Layer of resilience analysis (LORA)

the consequence is related to safety functions. It is possible to have multiple resiliency layers to mitigate the same resiliency demand. Each station should define its target resiliency performance to be used as a constraint to optimize the resiliency design of the station. Different resiliency layers are designed and quantified based on resiliency demand and station design and operation specifications.

10.5   Emergency Analysis of Charging Stations

Emergencies are possible consequences of a number of reasons, such as extreme weather conditions, personal emergencies, city emergencies, road emergencies, vehicle emergencies, or national emergencies [19]. The definition of emergencies will be used to define possible conditions and requirements for charging during these emergencies. Emergencies will impact the charging station at different levels due to changes in process variables associated with demand. Emergencies are also linked with resiliency demands. The emergency analysis of fast-charging infrastructure is presented in Fig. 10.7.

Emergency levels include world, country, state/province, city, area, vehicle, and human. Emergency nature is classified based on extreme weather (snow, rain, storm, temperature), health (pandemic and radiation), food, water, flood, terror, and cybersecurity. In order to manage these emergencies, an emergency index (EI) is associated with each level, including vehicle (EI-V), driver (EI-D), charging station (EI-C), and grid and energy system (EI-G). EI is also defined for other factors such as environment (EI-E), road (EI-R), traffic (EI-T), and service provider (EI-S). EI is linked with potential outcomes for each subsystem. Emergencies at the grid and

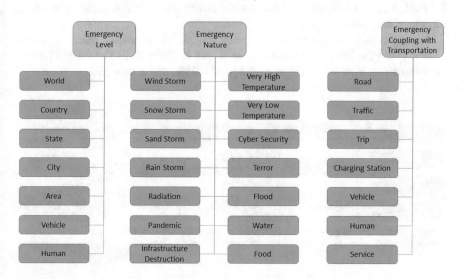

Fig. 10.7 Emergency analysis for charging infrastructure

energy system level will impact the energy supply to the station, which might cause no energy supply, reduced supply, or accidents. Similarly, emergencies at the station level will lead to no charging, reduced charging, or accidents. The vehicle level's emergency might lead to no move, reduced move, or accident. Emergencies at the driver level might lead to no driving, reduced driving, or accidents. The emergency is mainly an external factor that impacts each entity. Figure 10.8 shows emergency index analysis for the fast-charging station. Each emergency index is classified in five major levels: 5, "very high"; 4, "high"; 3, "medium"; 2, "low"; and 1, "very low." The emergency index associated with the environment, road, traffic, and service provider is specified by time interrupted, the area damaged, vehicles interrupted, human impacted, and trips impacted.

Figure 10.9 shows the analysis of emergency indexes of incoming vehicles for charging. EI-C1 is the emergency index for incoming vehicle-1. Similarly, EI-C2 and EI-C3 are total emergency indexes for incoming vehicle-2 and vehicle-3, respectively. EI-C includes vehicle, human trip, road, and traffic emergencies. These factors are linked with driver and environmental factors.

The charging station will evaluate the emergency index of the incoming vehicles as per Eq. (10.2):

$$EI-C(EV): \frac{(EI-V*v\_w)+(EI-D*d\_w)+(EI-R*r\_w)+(EI-E*e\_w)}{100} \quad (10.2)$$

The total emergency index has a value between [0,100] that represents the incoming vehicle, including vehicle, driver, the road for the outgoing trip, and environmental factors.

The weight factors are between [0,100] and are defined as follows: v_w is a weight factor of vehicle, d_w is a weight factor for the driver, r_w is a weight factor for road, and e_w is a weight factor for the environment, which includes weather, air, water, and public health.

To understand the proposed emergency model, let us assume three incoming vehicles with the associated attributes, as shown in Table 10.2. The weight factors are considered: v_w = 20%, d_w = 30%, r_w = 25%, and e_w = 25%.

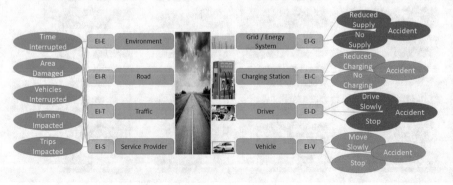

Fig. 10.8 Emergency index analysis for charging station

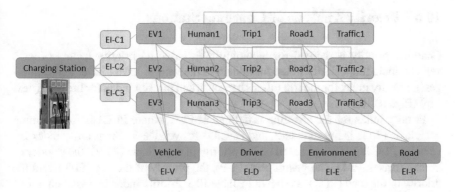

Fig. 10.9 Emergency index analysis for fast charging

Table 10.2 Example of emergency model estimation

EV-1	EI-V1	4
	EI-D1	4
	EI-R1	2
	EI-E1	1
EI-C1		$$\frac{\left(\left(4*20\right)+\left(4*30\right)+\left(2*25\right)+\left(1*25\right)\right)}{100}$$ $$= 2.75$$
EV-2	EI-V2	2
	EI-D2	4
	EI-R2	3
	EI-E2	4
EI-C2		$$\frac{\left(\left(2*20\right)+\left(4*30\right)+\left(3*25\right)+\left(4*25\right)\right)}{100}$$ $$= 3.35$$
EV-3	EI-V3	1
	EI-D3	2
	EI-R3	4
	EI-E3	4
EI-C3		$$\frac{\left(\left(1*20\right)+\left(2*30\right)+\left(4*25\right)+\left(4*25\right)\right)}{100}$$ $$= 2.80$$

The higher the charging emergency index, the higher the priority in the queue of incoming vehicles to charge. EV2 has the highest emergency index, which means it will be leading in the charging process. The total priority index includes the emergency index as well as other important indexes, as explained in the following section.

10.6   Priority Analysis of Charging Stations

Charging priority is defined for incoming EVs based on priority factors of each vehicle, including priority index for human in the vehicle level (driver and passengers), priority index for the trip (after charging), and priority index for the road (next trip) (Fig. 10.10).

Priority index for the charging is defined as PI-C, where PI-C1 is for incoming vehicle-1, PI-C2 is for vehicle-2, and PI-C3 is for vehicle-3. The priority index for vehicles PI-C is estimated based on the priority of the vehicle (PI-V), the priority of human (driver, PI-D, and passengers, PI-P), the priority of the trip (PI-T), and the priority of the road (PI-R), as shown in Table 10.3. Priority index is classified in five major levels: 5, "very high"; 4, "high"; 3, "medium"; 2, "low"; and 1, "very low." The human priority is estimated, as shown in Eq. (10.2), based on the average of driver's priority and passengers' priority using weight factors for the driver (pvd_w) and passengers (pvp_w), considering the number of passengers: n. The human priority index has a value between [1, 5], as shown in Eq. (10.3):

$$PI-H = \frac{(PI-D*pvd-w)}{(n+1)*100} + \frac{(n)*(PI-P*pvp\_w)}{(n+1)*100} \tag{10.3}$$

The total priority index is estimated based on weights of vehicle (pv_w), human (ph_w), trip (pt_w), and road (pr_w). Emergency index is integrated with the priority index based on emergency weight factor (ei_w).

An example is used to explain the estimation of the priority index of three incoming vehicles using the proposed priority model. The attributes for these vehicles are shown in Table 10.4. The weight factors are considered: pv_w = 30%, ph_w = 10%, pt_w = 25%, and pr_w = 25%. Moreover, the integration of the emergency index is based on

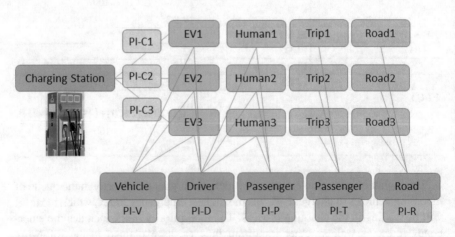

Fig. 10.10 Priority analysis of charging station

weight factor ei_w = 10%. The driver's priority is defined as pvd_w = 50%, and the priority of the passengers is pvp_w = 50%.

Equation 10.4 is used to estimate the total priority index for charging incoming vehicle:

Table 10.3 Example of emergency model estimation

PV-1	n	4
	PI-V1	4
	PI-D1	1
	PI-P1	4
	PI-H1	$$\frac{(1*50)}{(4+1)*100} + \frac{(4)*(4*50)}{(4+1)*100} = 0.1 + 1.6 = 1.7$$
	PI-T1	1
	PI-R1	3
	EI-1	2.75
PI-C1		$$\frac{(4*30)+(1.7*10)+(3*25)+(1*25)+(2.75*10)}{100} = 2.65$$
PV-2	PI-V2	2
	PI-D2	4
	PI-P2	4
	PI-H2	$$\frac{(4*50)}{(4+1)*100} + \frac{(4)*(4*50)}{(4+1)*100} = 0.4 + 1.6 = 2.0$$
	PI-R2	3
	PI-T2	4
	EI-2	3.35
PI-C2		$$\frac{(2*30)+(2.0*10)+(3*25)+(4*25)+(3.35*10)}{100} = 2.89$$
PV-3	PI-V3	1
	PI-D3	3
	PI-P1	2
	PI-H1	$$\frac{(3*50)}{(4+1)*100} + \frac{(4)*(2*50)}{(4+1)*100} = 0.3 + 0.8 = 1.1$$
	PI-R3	4
	PI-T3	2
	EI-3	2.80
PI-C3		$$\frac{(1*30+(1.1*10)+(2*25)+(4*25)+(2.80*10)}{100} = 2.19$$

$$PI - C(EV): \frac{(PI - V * pv\_w) + (PI - H * ph\_w) + (PI - R * pr\_w) + (PI - T * pt\_w) + (EI - C * ei\_w)}{100} \quad (10.4)$$

The estimated priority shows vehicle-2 as the higher priority to be charged first. Vehicle-1 is the second highest vehicle.

10.7   Vehicle Energy Management in Emergencies

EV, EB, ET, and EM will have local energy management system that will communicate with energy management system at the charging station. The functions covered by the vehicle energy management system include monitoring battery (or ultracapacitor) condition, SoC, and other health condition parameters such as temperature. Vehicle energy management will receive messages from nearest charging stations, and based on emergency indexes and priority indexes, the best option for charging station will be decided.

In order to understand the control strategy and energy management in each vehicle during emergencies, an example is selected as shown in Fig. 10.11. A number of nodes, which represent intersection points, are defined to link roads between EVs and charging stations.

The selected model parameters are defined in Table 10.4. Road index is used to quantify the road condition between nodes that will include trip delays (at a given time), number of lanes, and max capacity (Table 10.5).

In case of emergencies, a given area might be closed or impacted where vehicles will be able to access that area. Figure 10.12 shows the impacted area, marked in yellow, where vehicles' energy management system will select routes and charging stations outside this area.

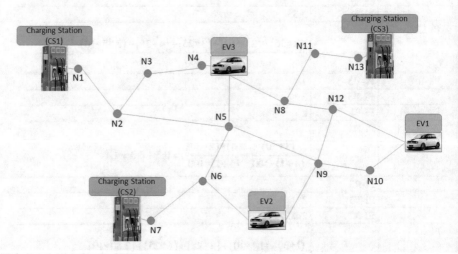

Fig. 10.11 Example of EVs and charging stations, with nodes, and roads

Table 10.4 Vehicle energy management model parameters

Charging station CS1	Charging capacity	CC1
	Charging units	CU1
	Max charging queue length	MQ1
	Charging load profile	CL1
	Emergency index	EI-C1
Charging station CS2	Charging capacity	CC2
	Max charging vehicles	CU2
	Max charging queue length	MQ2
	Charging load profile	CL2
	Emergency index	EI-C2
Charging station CS3	Charging capacity	CC3
	Max charging vehicles	CU3
	Max charging queue length	MQ3
	Charging load profile	CL3
	Emergency index	EI-C3
EV1	State of charge	SOC1
	Remaining distance	RD1
	Trip range	TR1
	Emergency index	EI-V1
	Distance to Station-1	DS11
	Distance to Station-2	DS12
	Distance to Station-3	DS13
	Road index to Station-1	RI21
	Road index to Station-2	RI22
	Road index to Station-3	RI23
EV2	State of charge	SoC2
	Remaining distance	RD2
	Trip range	TR2
	Emergency index	EI-V2
	Distance to Station-1	DS21
	Distance to Station-2	DS22
	Distance to Station-3	DS23
	Road index to Station-1	RI21
	Road index to Station-2	RI22
	Road index to Station-3	RI23
EV3	State of charge	SoC3
	Remaining distance	RD3
	Trip range	TR3
	Emergency index	EI-V3
	Distance to Station-1	DS31
	Distance to Station-2	DS32
	Distance to Station-3	DS33
	Road index to Station-1	RI31
	Road index to Station-2	RI32
	Road index to Station-3	RI33

(continued)

Table 10.4 (continued)

N1-N2	Road index	RI1-2
	Traffic delay	TD11
	Number of vehicles	EV12

Table 10.5 Constraints for energy management strategies in emergencies

Total incoming vehicles ≤ charging units + number of vehicles waiting in the queue
Total energy demand for all charging vehicles ≤ charging capacity
Total distance to Station-1 ≤ remaining distance
Number of vehicles in the road (N1N2) ≤ road capacity

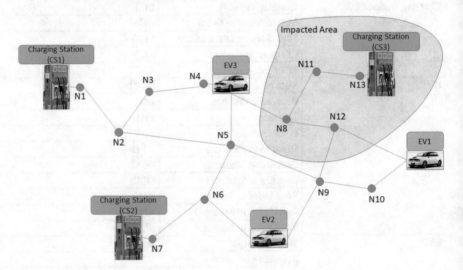

Fig. 10.12 Impacted area in emergencies

Each vehicle management will estimate the best charging station based on the defined constraints and optimization objective functions. The optimization algorithm for each vehicle includes road condition, vehicle condition, charging station condition, and environmental condition. The emergency index, road index, and priority index are estimated and used to prioritize charging of incoming vehicles. Figure 10.13 shows the framework for vehicle charging in emergencies.

10.8   Summary

This chapter presented discussions on harsh environments and emergency conditions. Different scenarios are presented and related to resiliency design. An integrated resiliency analysis framework, called layer of resilience analysis (LORA),

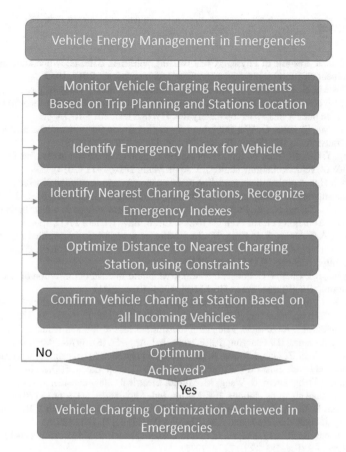

Fig. 10.13 Vehicle charging during emergencies

is presented using independent resiliency layers (IRLs). Different causes and related factors are discussed for the harsh environment in different levels, including worldwide, continent, country, state/province, city, area, building, and vehicle. The harsh environments due to different severe weather conditions are discussed in different levels and are linked with human and vehicle conditions as part of the charging process. Emergency scenarios are analyzed using different charging scenarios. Emergency indexes are introduced and used to assess different harsh conditions and evaluate the charging of incoming vehicles to the station under different conditions. Priority indexes are used to analyze priority to charge incoming vehicles with the considerations of emergency conditions at different levels: station, vehicle, driver, passengers, roads, trips, and environmental factors. Case studies are used to illustrate the proposed framework for charging in harsh environments and emergencies.

References

1. V.K. Khanna, Institute of Physics (Great Britain), publisher, Extreme-temperature and harsh-environment electronics: physics, technology and applications, 2017, 1952
2. A. Aldossary, S. Mahmoud, R. AL-Dadah, Technical feasibility study of passive and active cooling for concentrator PV in harsh environment. Appl. Therm. Eng. **100**, 490–500 (2016)
3. Lithium ion battery causes laptop battery fire, UWIRE text, 2015-10-28, p.1
4. S. Li, H. Ma, T. Saha, G. Wu, Bayesian information fusion for probabilistic health index of power transformer. IET Gener. Transm. Distrib. **12**(2), 279–287 (2018)
5. J. Wang, Y. Cui, M. Zhu, W. Sun, Probabilistic harmonic calculation in distribution networks with electric vehicle charging stations. J. Appl. Math. **2014**, 1–11 (2014)
6. S. Kim, J. Hur, A probabilistic modeling based on monte carlo simulation of wind powered EV charging stations for steady-states security analysis. Energies (Basel) **13**(20), 1 (2020)
7. W. Chen, L. Zhang, X. Pei, Probability evaluation of excess voltage in a distribution network with uneven charging electric vehicle load. J. Electr. Eng. Technol **17**(1), 15–23
8. H. Wang, A.F. Abdin, Y.-P. Fang, E. Zio, Resilience assessment of electrified road networks subject to charging station failures. Comput-Aided. Civ. Infrastruct. Eng. **37**(3), 300–316 (2022). https://doi.org/10.1111/mice.12736
9. I. Aravena, S.J. Chapin, C. Ponce, Decentralized failure-tolerant optimization of electric vehicle charging. IEEE Trans. Smart Grid **12**(5), 4068–4078 (2021)
10. X. Yu, K. Qian, H. Gao, Design of safe operation scheme of charging station based on AI particle swarm optimization, IOP conference series. Mater. Sci. Eng. **631**(5), 52026 (2019)
11. M. Soleimani, M. Kezunovic, Mitigating transformer loss of life and reducing the hazard of failure by the smart EV charging. IEEE Trans. Ind. Appl. **56 (5)**, 5974–5983 (2020)
12. D. Mao, C. Yuan, Z. Gao, J. Wang, R. Zhao, Online prediction for transmission cascading outages induced by ultrafast PEV charging. IEEE Trans. Transp. Electr. **5**(4), 1124–1133 (2019)
13. B. Wang, P. Dehghanian, S. Wang, M. Mitolo, Electrical safety considerations in large-scale electric vehicle charging stations. IEEE Trans. Ind. Appl. **55**(6), 6603–6612 (2019)
14. H.-S. Han, E. Oh, S.-Y. Son, Study on EV charging peak reduction with V2G utilizing idle charging stations: the Jeju Island case. Energies (Basel) **11**(7), 1651 (2018)
15. H.A. Gabbar, Resiliency analysis of hybrid energy systems within interconnected infrastructures. Energies (Basel) **14**(22), 7499 (2021)
16. K.-W. Hu, C.-M. Liaw, On a bidirectional adapter with G2B charging and B2X emergency discharging functions. IEEE Trans. Ind. Electron **61**(1), 243–257 (1914)
17. R. Moore, Plug-in vehicle charging stations: procedures for dealing with various charging station emergencies. Firehouse **41**(6), 34 (2016)
18. Z. Li, W. Tang, X. Lian, X. Chen, W. Zhang, T. Qian, A resilience-oriented two-stage recovery method for power distribution system considering transportation network. Int. J. Electr. Power Energy Syst. **135**, 107497 (2022)
19. J.P. Minas, N.C. Simpson, Z.Y. Tacheva, Modeling emergency response operations: a theory building survey. Comput. Oper. Res. **119**, 104921–104918 (2020)

Chapter 11
Autonomous Transportation

Hossam A. Gabbar

11.1 Autonomous Transportation

As urbanization increases—an additional 2.5 billion people will live in cities by 2050—cities and suburbs will undergo significant transformations to create sustainable living conditions for their residents. Energy and transportation are the twin pillars of these transformations, and both will require radical adaptation to meet the demographic and economic growth without increasing congestion and pollution.

Electric mobility is widely seen today to improve air quality and meet climate goals, and it is by nature integrated in a comprehensive vision for smarter cities. As electric vehicles (EVs) become more affordable, it is predicted that they will constitute almost a third of new car sales by the end of the next decade. Ride-sharing continues to surge, with estimates that by 2030, it will account for more than 25% of all miles driven globally, up from 4% today. Electrified transportation can be used as a decentralized energy resource and provide new, controllable storage capacity and electricity supply that is useful for the stability of the energy system. In markets where regulation allows EVs to be used as a source of flexibility, energy players start betting on this vision, with cars working as "batteries on wheels."

These changes are just the first hints of what is to come as we will soon see connected and autonomous vehicles (CAVs) and commercial fleets of EVs integrated as parts of everyday life. In the future, CAVs will also cost significantly less per mile than vehicles with internal combustion engines for personal use—by as much as 40%—and could also reduce congestion and traffic incidents. Autonomous and

---

This chapter is contributed by Hossam A.Gabbar.

H. A. Gabbar (✉)
Energy Systems and Nuclear Science, University of Ontario Institute of Technology,
Oshawa, ON, Canada
e-mail: Hossam.gabbar@uoit.ca

© Springer Nature Switzerland AG 2022
H. A. Gabbar, *Fast Charging and Resilient Transportation Infrastructures in Smart Cities*, https://doi.org/10.1007/978-3-031-09500-9_11

connected vehicle technology is expected to transform the nation's transportation system over the coming decades, with significant implications for the planning and design of cities and regions. Autonomous vehicles (AVs), also known as driverless or self-driving cars, have experimented with city streets for several years. This technology is moving very quickly, with the 11 largest automakers planning to have fully autonomous vehicles on highways between 2018 and 2021 (arriving somewhat later in urban driving conditions). AV technology, as defined by the International Society of Automotive Engineers, ranges from a baseline of no automation up to five levels of increasing autonomy:

- Level 1, driver assistance (e.g., adaptive cruise control)
- Level 2, partial automation (e.g., Tesla's autopilot)
- Level 3, conditional automation (e.g., human drivers serve as backup for an autonomous system that operates under certain conditions)
- Level 4, high automation (e.g., Google/Waymo test cars)
- Level 5, full automation (e.g., no steering wheel in the vehicle)

Researchers estimate that about two-thirds of vehicles in Canada currently have some connectivity (i.e., embedded telematics); by 2022, approximately 70%–95% of new cars in Canada will have vehicle-to-infrastructure (V2I), vehicle-to-vehicle (V2V), and other telecommunications capabilities (e.g., vehicle to smartphone). Currently, vehicles available to consumers are primarily Level 1 or 2 automation. While a number of major manufacturers plan to launch autonomous passenger cars in the next year, the consensus in the literature is that deployment of Level 4 or 5 vehicles on public roads will not be commonplace until the 2030s or 2040s. In fact, major auto-manufacturers, including Ford and GM, have indicated that their first autonomous fleets will be dedicated to commercial operations.

Autonomous cars use artificial intelligence (AI) systems to interact with their surroundings and travel safely. One form of artificial intelligence being used in fully autonomous cars is called fuzzy logic. Fuzzy logic allows the computer to use approximate reasoning instead of fixed reasoning to interpret situations the car may encounter while traveling on public roadways. This means that the car interprets situations as something other than being entirely true or completely false. Another form of artificial intelligence being used in fully autonomous cars is called swarm intelligence. The definition of swarm intelligence is "the collective behavior of decentralized, self-organized systems, natural, or artificial." In common terms, swarm intelligence traces the velocity and period it takes groups of particles to return to the computer from which they were transmitted. Using a mathematical relationship, the computer can then determine the location of its boundaries and other structures within the environment. Light detection and ranging (LIDAR) is another method the autonomous car uses to map out its surroundings instantaneously during travel. LIDAR operates the same way a radar system would, but it transmits laser lights instead of transmitting radio waves. Stefano Young from

Nature Photonics says, "The system at the core of the LIDAR technology is a sensor called the HDL-64E that uses 64 spinning lasers and accumulates 1.3 million points per second in order to reconstruct a virtual map of its surroundings." These virtual maps created by LIDAR and current GPS maps allow the autonomous car to construct highly accurate depictions of its surrounding environment.

Major changes to the transportation system are expected to occur as the development and adoption of connected and autonomous vehicles (CVs/AVs) continues. Notably, passengers' safety [1, 2] and transport efficiency are expected to increase. According to the National US Highway Traffic Safety Administration (NHTSA), over 94% of collisions are caused by human error. Connected and autonomous vehicles have the potential to eliminate the element of human error, likely resulting in increased overall passenger safety. Heightened V2V, V2I, and V2X connectivity may also decrease the risk of collisions as vehicles become more aware of and reactive to their surroundings (e.g., other drivers' speed, traffic signals, and impending collisions). As adoption of ride-sourcing programs such as Uber and Lyft increases, in parallel with potential shifts toward platform-owned autonomous vehicle fleets, experts believe that the demand for personal vehicle ownership and potentially the number of vehicles on the road will decline. That said, the shift toward car-sharing and ride-sourcing may result in a range of scenarios. Other experts predict that the rise of car-sharing/ride-sourcing may increase road congestion as passengers who typically use modes of public transportation or are unable to drive (due to disability) begin using car-sharing and ride-sourcing programs. According to a study conducted by the Insurance Institute of Canada, these CV/AV trends will also significantly affect the commercial trucking/shipping sector, mainly as autonomous technologies will likely first be adopted by this sector. Figure 11.1 shows the main research areas connected with CAV. Figure 11.2 shows a framework for KPI-based simulation.

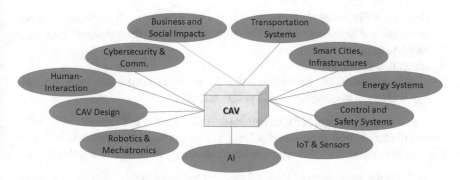

Fig. 11.1 CAV main research areas

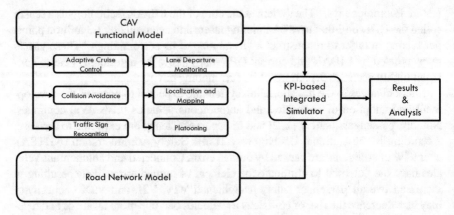

Fig. 11.2 Connected and autonomous vehicle functional model and KPI-based simulation

11.2 Charging Requirements for Autonomous Transportation

In order to implement effective charging infrastructure for autonomous transportation, the following main tasks are considered:

Task 1 Functional modeling of connected, autonomous vehicles, and performance metrics:

1. Road network model with potential charging points: the description of the road network, the definition of the properties of an intersection, and the definition of traffic rules [3]
2. Mobility/traffic model of autonomous and connected electric vehicles to optimize charging infrastructure: integrating charging process with traffic and mobility models
3. Defining KPIs for charging infrastructures of autonomous transportation as per technical and legislative (traffic standards, law of land, federal regulations, etc.) requirements
4. Integrated simulation [4]

    Application: It can be used as a tool for research on new standards for intelligent transportation systems and public safety.

Task 2 Analysis, design, and requirement specifications of IoT (Internet of Things), communications, and cybersecurity layer [5] of connected autonomous vehicles (CAVs) as integrated with charging infrastructures, components, and technologies:

1. Technology assessment of CAV subsystems and associated charging subsystems and electronic control units (ECUs)
2. Assessment of intra-vehicular communication technology for V2V charging and industry-standard protocols and state-of-the-art R&D (research and development) tools
3. Assessment of industry-standard protocols and state-of-the-art R&D tools for V2X communications [6] with considering charging infrastructures

4. Assessment of industry-standard and state-of-the-art R&D tools for cybersecurity testing of CAVs as integrated with charging infrastructures [7]
5. Assessment of IoT standards and R&D tools for intelligent transportation systems and charging infrastructures

Application: It can be used as a state-of-the-art knowledge base for research and development.

Task 3 Control system architecture, spec, and features in CAV:

1. Assessment of CAV surroundings' mapping technology to interface with charging infrastructures (sensors and algorithms for localization, environment perception, sensor data fusion, and environment prediction)
2. Assessment of autonomous driving algorithms [8] and include charging profiles (e.g., motion control, battery charge control, powertrain control [9, 10], eco-driving)
3. Assessment of state-of-the-art R&D tools for CAV control architecture testing as integrated with charging infrastructures [11–13]

Application: It can be used as a state-of-the-art knowledge base for research and development. Figure 11.3 shows the possible architecture of CAV with the control system.

Task 4 Analysis of safety and protection systems, functions, and risk analysis of CAV with charging process:

1. Assessment of current automotive industry standards for CAV functional safety [14] and risk management (e.g., automotive safety integrity levels (ASILs) as per ISO 26262 (functional safety)), as shown in Fig. 11.4

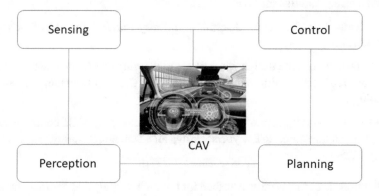

Fig. 11.3 Autonomous control architecture of connected and autonomous electric vehicle

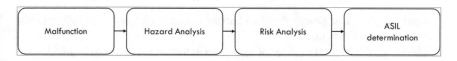

Fig. 11.4 Automotive safety integrity level (ASIL) determination workflow

2. Design of novel functional safety analysis framework for CAVs with charging systems

Applications:

- It can be used as a tool for research on new standards for intelligent transportation systems and public safety.
- It can be used as a state-of-the-art knowledge base for research.

Task 5 CAV technology assessment, analysis, R&D, and innovations:

1. Technology assessment of CAV subsystems and charging systems associated with electronic control units (ECUs)
2. Assessment of state-of-the-art R&D and innovations due to CAV applications in smart grid, smart cities, CAV as energy storage options, blockchain, and transactive energy

Application: It can be used as a state-of-the-art knowledge base on CAV technology for research.

Task 6 Analysis of national and international standards related to CAV and charging systems:

1. Assessment of intra-vehicular communication industry standard protocols
2. Assessment of industry standard protocols for V2X communication with interfaces to charging systems
3. Assessment of industry standard for cybersecurity testing of CAV charging systems
4. Assessment of IoT intelligent transportation systems
5. Assessment of current automotive industry standards for CAV charging functional safety and risk management [15]

Application: CSA can determine priority areas for CAV charging standardization.

Task 7 Define possible design specs of CAV charging test facility with operation scenarios [16] for testing and integration of CAV auto-manufacturers and technologies:

1. Design of a single integrated testbed for CAV charging, V2X communication, control architecture, IoT, cybersecurity, and functional safety, as shown in Fig. 11.5

Application It can be used as a testbed for research, testing, and certification activities of CAV charging systems, components, and functional safety.

There are potential management challenges to deploy CAVs, as below:

- Demographic: The geographical and socioeconomic elements that influence the composition of a population
- Market size: The total number of cars, both autonomous and conventional
- External environment: Road network and physical environment, as well as their CAV readiness

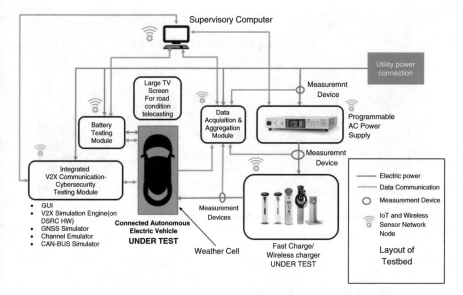

Fig. 11.5   Design of an integrated test platform for CAV charging

- Perception/adoption rate: The population's attitude or view of CAVs, which predicts future acceptance
- Regulations: Legislation enacted in a particular market to demand the ownership or sale of CAVs, as well as the creation of necessary infrastructure
- Willingness to pay (WTP): A cost-benefit analysis of whether the population believes the associated benefits of acquiring CAVs outweighing the financial expense

There are potential challenges related to the deployment of CAVs, as below:

- Recurrence and control of congestion
- Safety of pedestrians and cyclists
- Predictability of travel times
- Intersection management and operations
- Emergency response, incidents, and evacuations
- CAV effectiveness in a mixed-vehicle setting

These challenges are reflected in charging models of CAVs, which are applied on the selected case study in the following section.

11.3   Case Study

This section will discuss a case study of charging infrastructure for CAVs [17]. The first step is understanding the mobility models of CAVs in a given region, which will be the basis for determining the charging load profiles for CAV.

The annual number of passengers traveling through the Turin-Caselle Airport from the year 2010 to the year 2019 (in millions) is represented below [18], as

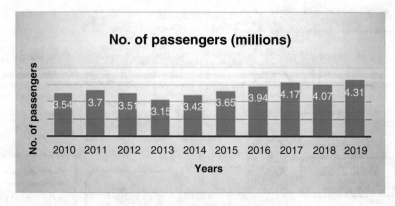

Fig. 11.6 Annual number of passengers at the Turin Airport

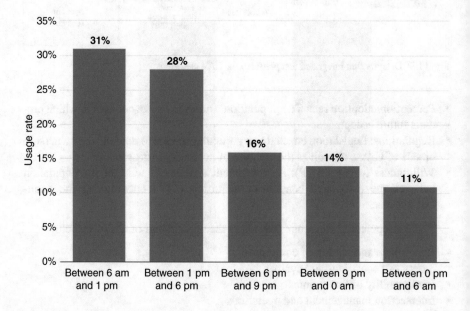

Fig. 11.7 Car-sharing usage in Turin City

shown in Fig. 11.6. The bar graph is derived from the report of the Airport Traffic APT given by the Traffic data Italia. This data helps us create a profile for mobility demand for our scenarios and other necessary calculations [19]. The car-sharing example in Turin is shown in Fig. 11.7.

CAV planning is based on route design, as shown in the example in Fig. 11.8.

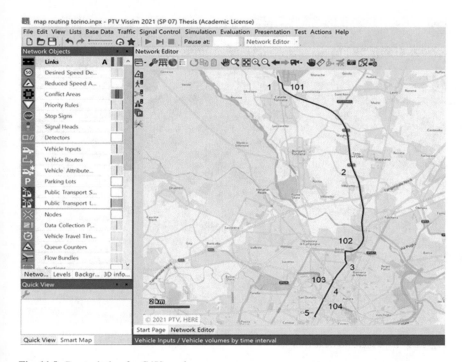

Fig. 11.8 Route design for CAV services

11.4   Base Scenario

Two separate lanes are designed for the base scenario for CAV mass transit purposes, that is, car-sharing in our case. One dedicated lane will be used for travelling from Turin Airport to Turin City Center. At the same time, the other will be dedicated for the transit from the city center toward the airport, as shown in Fig. 11.9 [17]. Only CAVs would be allowed to travel on the dedicated lanes in the scenarios planned, and the performance indicators are valued and presented in the simulation and results chapter. The scenarios were modeled on the following consideration, supported by the statistical data provided in the literature [20]:

- Yearly passengers at Turin Airport are approximately provided in the data as 950,000. These are segregated as passengers at arrival and departures.
- This depicts the daily passengers traveling to and from the airport, more or less at 2600.
- From the statistical data and the mobility demand of car-sharing users in Turin City [18], passengers inbound and outbound from the airport are approximated at 30% of the daily demand, that is, 780 passengers.
- Peak hours of travel for car-sharing service usage are depicted in the statistical data as 6 hours (7 am–1 pm).

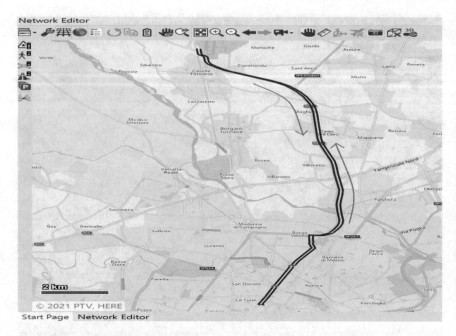

Fig. 11.9 Base scenario route design

- Also, the daily demand of the service users during peak hours is about 31% of daily passengers, that is, approximated at 240 passengers.
- Load factor is considered one because maximum utilization is possible and thus assumed.
- From the literature, 10% of the actual fleet size calculated must be considered and added as the reserved fleet in the total fleet size.
- The following formula is used for the fleet size calculation [20] as supported in the literature: fleet **size** = (*Max. Load* ∗ *Cycle Time*) ÷ (*Vehicle Capacity* ∗ *Load Factor* ∗ *Peak hours*).

11.5   The Scenario of Fixed Pick-Up and Drop-Off Points

A fixed pick-up and drop-off points are allocated and set in this scenario, as shown in Fig. 11.10. The scenario planned would imply that only two starting/ending points of the journey would be present: the airport parking and the parking lot dedicated at the city center for CAV parking. CAV charging is analyzed in view of these scenarios while considering service points, mobility demands, and charging infrastructures.

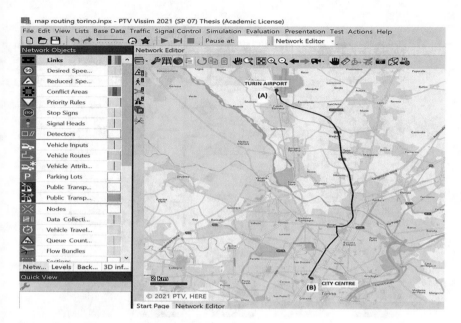

Fig. 11.10 Scenario-A design of fixed pick-up and drop-off

11.6 Mapping CAV Routes to Charging Infrastructure

The number of CAVs in each road segment will be translated into charging load at each point, considering battery status at the charging point. The analysis will consider parameters related to battery condition such as type of vehicle (EV, e-bus, e-truck, CAV), battery size and type, state of charge (SoC), priority index for each vehicle, battery temperature, and the number of vehicles in the queue [20, 21]. Only chargers with wireless charging will be used with CAV. The charging infrastructure planning will consider the penetration ratio of CAV versus conventional vehicles. Figure 11.11 shows design of fast-charging station as integrated with microgrid to support transportation electrification of EVs, e-buses, e-trucks, CAVs, and marine.

The integrated planning of charging infrastructures includes CAV and other electric transportation systems, as shown in Fig. 11.12. The integrated planning is configured and optimized with AI algorithms in view of the defined vehicle, infrastructures, and mobility parameters while considering other factors such as social, economic, health, environmental, and safety.

11.7 Summary

With the world moving toward autonomous transportation, there are needs to plan to charge infrastructures to support CAV by providing wireless charging stations. The planning will include the type, size, and location of charging stations. The

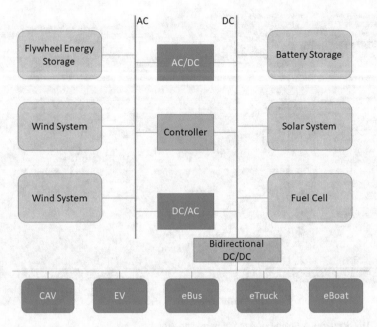

Fig. 11.11 Fast-charging station with microgrid for transportation electrification

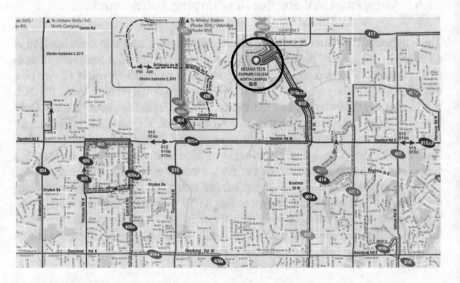

Fig. 11.12 Mapping charging stations in selected region

analysis of autonomous transportation is performed to enable proper planning of the supporting charging infrastructure.

References

1. K. Kockelman, P. Avery, P. Bansal, S.D. Boyles, P. Bujanovic, T. Choudhary, L. Clements, G. Domnenko, D. Fagnant, J. Helsel, R. Hutchinson, M. Levin, J. Li, T. Li, L. LoftusOtway, A. Nichols, M. Simoni, D. Stewart, Implications of connected and automated vehicles on the safety and operations of roadway networks. Fhwa/Tx-16/0-6849-1 **7** (2016)
2. M. Singh, S. Kim, Safety Requirement Specifications for Connected Vehicles (2017)
3. L. Ye, T. Yamamoto, Modeling connected and autonomous vehicles in heterogeneous traffic flow. Phys. A Stat. Mech. its Appl. **490**, 269–277 (2018)
4. S. Hallerbach, Y. Xia, U. Eberle, F. Koester, Simulation-based Identification of Critical Scenarios for Cooperative and Automated Vehicles, no. April, 2018
5. P. Wooderson, D. Ward, Cybersecurity Testing and Validation, SAE Tech. Pap. (2017)
6. I. Ivanov, C. Maple, T. Watson, S. Lee, Cyber Security Standards and Issues in V2X Communications for Internet of Vehicles, pp. 1–6 (2018)
7. X. Zheng, L. Pan, H. Chen, R. Di Pietro, L. Batten, A testbed for security analysis of modern vehicle systems, in Proc. – 16th IEEE Int. Conf. Trust. Secur. Priv. Comput. Commun. 11th IEEE Int. Conf. Big Data Sci. Eng. 14th IEEE Int. Conf. Embed. Softw. Syst. Trust. 2017, no. October, pp. 1090–1095 (2017)
8. D. Gruyer, V. Magnier, K. Hamdi, L. Claussmann, O. Orfila, A. Rakotonirainy, Perception, information processing and modeling: critical stages for autonomous driving applications. Annu. Rev. Control **44**, 323–341 (2017)
9. N. Kim, S. Choi, J. Jeong, R. Vijayagopal, K. Stutenberg, A. Rousseau, Vehicle level control analysis for Voltec powertrain. World Electr. Veh. J. **9**(2), 1–12 (2018)
10. F. Akkaya, W. Klos, T. Schwämmle, G. Haffke, H.C. Reuss, Holistic testing strategies for electrified vehicle powertrains in product development process. World Electr. Veh. J. **9**(1), 1–10 (2018)
11. W. Way, C. City, Autonomous Vehicle Testing Registry Application, vol. 326, pp. 1-9 (2018)
12. P. Koopman, M. Wagner, Challenges in autonomous vehicle testing and validation. SAE Int. J. Transp. Saf. **4**(1), 2016-01-0128 (2016)
13. H.F. Prasetyo, A.S. Rohman, H. Hindersah, M.R.A.R. Santabudi, Implementation of Model Predictive Control (MPC) in electric vehicle testing simulator, in *Proceeding – 4th Int. Conf. Electr. Veh. Technol.* ICEVT 2017, vol. 2018-January, pp. 48–54 (2018)
14. M.S. Rahman, M. Abdel-Aty, L. Wang, J. Lee, Understanding the highway safety benefits of different approaches of connected vehicles in reduced visibility conditions. Transp. Res. Rec. (2018)
15. J.A. Ross, A. Murashkin, J.H. Liang, M. Antkiewicz, K. Czarnecki, Synthesis and exploration of multi-level, multi-perspective architectures of automotive embedded systems. Softw. Syst. Model. (2017)
16. S. Chhawri, G.R. Lane, S. Tarnutzer, T. Tasky, Smart vehicles, automotive cyber security & software safety applied to leader-follower (Lf) and autonomous convoy operations (Aco), Ndia Gr. Veh. Syst. Eng. Technol. Symp. (2017)
17. Muhammad Sami Ullah Khan, Master Thesis in Automotive Engineering, "Deployment of Connected Autonomous Vehicles (CAVs) for Mass Transit using Car Sharing Services," Collegio di Ingegneria Meccanica, Aerospaziale, dell'Autoveicolo e della Produzione

18. I.C.S. Project, Car Sharing in Italy," no. june, 2003
19. R.G. Mugion, M. Toni, L. Di Pietro, M.G. Pasca, M.F. Renzi, Understanding the antecedents of car sharing usage: an empirical study in Italy. Int. J. Qual. Serv. Sci. (2019)
20. L. Wright, H. Walter, The BRT Planning Guide. Itdp, 1076 (2017)
21. R. Fatima, A.A. Mir, A.K. Janjua, H.A. Khalid, Testing study of commercially available lithium-ion battery cell for electric vehicle. in Proc. – 2018, IEEE 1st Int. Conf. Power, Energy Smart Grid, ICPESG 2018, no. April, pp. 1–5 (2018)

Chapter 12
Transportation with Electric Wheel

Hossam A. Gabbar

12.1   Introduction

There are continuous developments and improvements to the transportation infra-structure to meet mobility and social demands. Electric vehicle (EV) is one major milestone in transportation technologies that led to a reduced GHG (greenhouse gas) emission. The recent improvements of battery energy storage systems enabled more extended range and improved energy efficiency and power system. There are other efforts to improve the powertrain and electric motor functionalities and per-formance and control systems. The complexity of EV power system and limitations of recovering maximum power from braking systems motivated researchers to see improved solutions. Among these solutions, electric wheel is introduced. Studies are made to improve motion coordination and powertrain control with a series-parallel hybrid vehicle with an electric wheel [1]. The vehicle's performance with electric wheel is optimized based on driver condition while modeling motor brake, active suspension systems, and steering system [2].

The analysis of EV and comparison with electric wheels should consider weight, safety, energy efficiency, stability, and lifecycle cost.

The in-wheel electric motor is also known as the electric wheel. The main struc-ture of the electric wheel is shown in Fig. 12.1.

Electric wheel requires the effective design of in-hub electric wheel motors. The improved design adds benefits to the performance of the existing electric vehicle. The introduction of individual wheel motors reduces complexity while still

---

This chapter is contributed by Hossam A.Gabbar.

H. A. Gabbar (✉)
Energy Systems and Nuclear Science, University of Ontario Institute of Technology,
Oshawa, ON, Canada
e-mail: Hossam.gabbar@uoit.ca

© Springer Nature Switzerland AG 2022
H. A. Gabbar, *Fast Charging and Resilient Transportation Infrastructures in Smart Cities*, https://doi.org/10.1007/978-3-031-09500-9_12

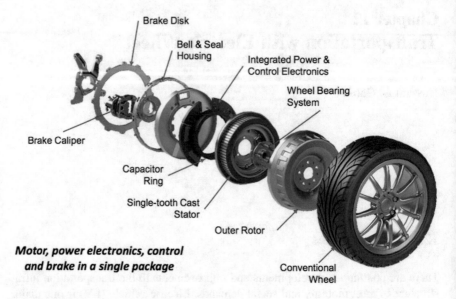

Fig. 12.1 Structure of in-wheel motor or electric wheel [46]

providing torque vectoring. Differential is a system of gears that allows different drive wheels on the same axle to rotate at different speeds, for example, during turn [3]. Differentials are required to split the torque should the vehicle has one wheel moving faster than the other. Differentials are typically very heavy and require a lot of expensive machining and maintenance. Due to friction, electric wheels will also improve interaction with roads [4]. As a mechanical interface system, the introduction of the differentials leads to energy losses as it has a lower efficiency than most electrical components. However, the mechanical system does offer higher reliability. The development of the electric wheel, including its motors, will provide more opportunities to enhance the overall reliability and improve driving torque. The transformation of internal combustion vehicles by introducing electric wheels will be a viable solution to reduce GHG emissions. In addition, electric wheels can be introduced in EVs to enhance driving torque, stability, and energy performance.

There are a number of vehicles that implemented in-wheel motor technology. Some companies are progressing with a similar motor design for each wheel. The motors are typically located inboard to make the design easy and accessibility simpler. A polish company called Rimac made electric supercars that utilize four-wheel technology. They progressed in research to enhance torque vectoring, which is reflected to the enhanced performance [5].

12.2   Regenerative Braking System

Tesla produced Roadster electric car and applied a number of experiments to improve energy performance. Tesla wants to improve the regenerative braking system to adjust to the user's preference and opinion on how much energy could be recovered. The electric regenerative braking is similar to the motor efficiency as the same motor works in reversed polarity. Tesla says that the motor is 80% efficient, which means that the braking efficiency of 0.8 multiplied by 0.8 results in a 64% efficiency. The regenerative braking is managed using the motor controller to reverse the polarity when braking and add power to the reverse direction to slow the vehicle down. However, this method means that there would be power being used to propel the motor and slow it down. This method relies a lot on the cooling on the motor since the input power in the reverse direction would generate a lot of heat and to a great extent would cause demagnetization of the motor [6].

Different types of regenerative systems utilize compressed air, which only requires the addition of air tanks and some other control valves to convert a standard internal combustion engine into a compressed air-operated engine with functioning regenerative braking. The new engine acts as a compressor converting the kinetic energy into compressed air stored in the air tanks. The solution offered a cheap alternative to manufacturers to introduce their cars to the hybrid market. The schematic for the pneumatic hybrid system is shown in Fig. 12.2, which resembles an internal combustion engine diagram except that it lacks the implementation of spark ignition. This means that the only functional component is the valve train, which allows the solenoid activated valve to release the air into the chamber to force the cylinder downward in order to rotate the crank [7].

The use of regenerative braking systems is not limited to production vehicles. One of the largest fuel consumers and producers of emissions is heavy-duty vehicles (HDVs), which require long-distance travel. A different hybrid system is

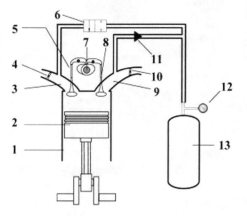

1. Cylinder
2. Piston
3. Intake Chamber
4. Intake Throttle Valve
5. Intake Valve
6. Electropneumatic Solenoid Valve
7. Cam Shaft
8. Exhaust Valve
9. Exhaust Chamber
10. Exhaust Throttle Valve
11. Check Valve
12. Pressure Sensor
13. Air Tank

Fig. 12.2   Schematic diagram of a pneumatic hybrid system [7]

available, which offers an electrically powered vehicle with an internal combustion engine range extender. However, this technology requires significant investments in the electrical drivetrain field. The hydraulic system has an improved performance over the standard compression ignition engine or an internal combustion engine. It provides a cheap conversion, which offers a simple integration for regenerative braking system applications. Unlike electric hybrid vehicles, HDVs are ideal for switching to a system that involves regenerative braking since the most significant difficulty with these vehicles is operating at low speeds due to their high mass. Figure 12.3 shows a schematic of the pneumatic/hydraulic hybrid system. An example of a 19 ton HDV shows that it loses 1.3–14.6MJ of energy depending on the stopping distance, which could be used to recirculate back into the energy storage system. In order to create a fast-operating regenerative system, hydraulic circuits combine pneumatics and hydraulics; however, for the accumulator, the increase in size is proportional to the amount of energy stored [8].

There are a number of parameters to be considered when designing a regenerative braking system. The coefficient of road adhesion is a significant factor in fuel economy and braking. Therefore, it is essential to consider that not all surfaces that vehicles operate on are the same, which means that the regenerative braking will not operate the same. After analyzing all road surfaces and gradients, it is fair to construct an algorithm that formulates the appropriate brake settings depending on the surface which the wheels of the vehicle travel over. These road conditions are typically responsible for 1–3% of the vehicle's fuel economy. The two main categories that regenerative braking can divide into are parallel braking and serial braking. In serial braking, there are advantages like coordinating regenerative and friction brakes, unlike parallel braking. Parallel control strategy features a simple implementation with additional installation components [9].

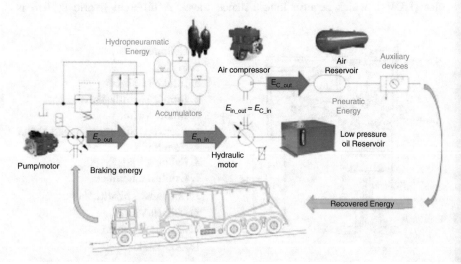

Fig. 12.3 HDV pneumatic/hydraulic hybrid system [8]

A safety feature that is fitted to most vehicles is an antilock braking system that prevents the wheels from locking, causing the car to skid uncontrollably. One idea of the regenerative system adaptations is to replace the mechanical passive antilock brakes with a brake by wire with a regeneration system to replace the abs feature but still achieve the same goal. Figure 12.4 shows a schematic of the regenerative braking system within EV. This can be done in several different strategies; first is when the motor gradually withdraws when the ABS is switched on. The second strategy sets the motor's torque to the maximum set value to apply as much braking torque as possible. However, this could result in the locking of the wheels. Finally, the last method is similar to the previous one since it will apply either the maximum brake torque with the motor or the reverse driving force on the motor is applied in order to slow the vehicle down [10].

Some applications to regenerative braking can include the use of adaptive cruise control, which can govern this feature when the user isn't driving; this system uses a motor, battery, and hydraulic system. The hydraulic system has a simulator that records the pressure required force for a specific deceleration. This can then be replicated by the pressurization control without input by the user while in adaptive cruise control. Unlike the typical energy storing systems used to reduce energy loss, this system takes the regenerative term to a different level of understanding by generating the user's required input [11].

Some types of hybrids have different algorithms which utilize the regenerative braking among the series-parallel hybrids. This process involves the use of a fuzzy logic controller, which is governed by a PID (proportional-integral-derivative) controller which communicates via a PWM (pulse width modulation) connection. Upon simulating the braking strength at a different speed, it is found that the braking torque of the motor increases with the increase in power to the motor. When the motor is switched to braking, the motor recovers energy all the time; however, toward the end, the battery capacity will increase in order to recover energy. Hydraulic brakes are also used as a constant to bring down the braking torque range.

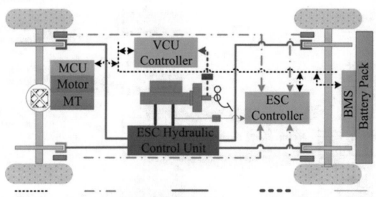

Can Bus   Wheel Speed   Hydraulic Pipeline   Displacement   Pressure

Fig. 12.4 Typical structure of an EV with a regenerative braking system [10]

After simulating the different speeds to evaluate this feature, the simulation found that the recovery rate is most significant when the vehicle initially achieves 60km/h [12].

Neglecting the power consumption of accessories, high amounts of energy can flow back into the energy storage system under ideal conditions. Figure 12.5 shows the energy flow of EV with and without a regenerative braking system. The standard configuration of the drivetrain for a front-wheel drive electric car can have this setup where the permanent magnet synchronous motor can work in two different types of settings, first which drives the motor and the second which uses the motor as a generator. The use of the generation mode will lead to efficiency of the energy usage and driving range extent by 11.18% and 12.58%, respectively. The studies on energy efficiencies show that regeneration will negatively impact efficiency. However, it is proven to significantly contribute to extending the vehicle's range. The proper design and testing will lead to improved efficiency, considering losses due to energy transformation from the motor [13].

A proposed technology for using a controllable regenerative braking system includes a circuit topology and decentralized active disturbance rejection controllers (ADRC). When experimenting with this design, it is found to brake with constant torque. The ultracapacitor and dissipative resistor act as the regenerative braking system. Where it stores, the high-speed energy is stored in the ultracapacitor and the resistor stores the low-speed energy. The ultracapacitor follows the law of induction due to electromagnetism to act as the regenerative braking system. This

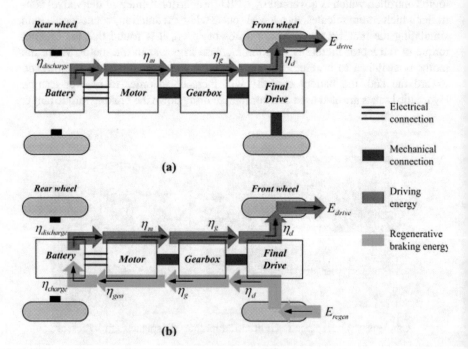

Fig. 12.5 The energy flow of the case study vehicle with and without a regenerative brake [13]

system, however, heavily relies on the dissipating resistor. The operational modes switch controller (OMSC) is used to work with the ADRC where it operates constrained within the ADRC using a decentralized set of coordinates [14].

12.3 Electric Wheel

Rolling road jig outlines the essential parameters involved with the rolling friction. This is vital since it helps to understand what forces are influenced when slipping or if they experience static contact. A better understanding of the force values will lead to a better analysis of the forces required to overcome and achieve the correct output. The article demonstrates that the increase of load on the wheels means there is a greater force to overcome. Here, the article provides the rolling friction coefficients for caster wheels used in our initial rolling road testing jig [5].

When considering the use of high-speed rotations, considerations must be taken to consider how to reduce as much energy lost in the process. This is very important since this is a testing jig that needs to measure as accurately as possible in the real scenario. Various types of bearings are used in the industry to different applications. The article explains the coefficients that are associated with each type. Furthermore, the calculation is shown to find the forces generated within the bearing. Other factors are discussed as being necessary to the design consideration when using bearings, such as the vibration frequency factors when the life of the bearing's usage is spent and approaches a frequency associated with impending failure. This is why there is regular maintenance in the industry and are a certain number of hour's usage [14].

For the calculations of the rolling resistance, the coefficient of rolling resistance will change depending on the surface and the type of material used for the wheel. This article combines the coefficient associated with types of vehicle wheels on various roads. However, when finding these coefficients, some parameters must remain constant. For rolling resistance, the tire pressure greatly affects the contact area, either increasing or decreasing the grip provided to the driving wheel [15].

When the assembly is constructed, measurements are required to determine specific parameters such as various forces, pressure, and torque values. In order to measure these, the sensors are made from insulated-based materials with a metal foil which deforms changing the resistance values. The strain gauge can then measure the change in resistance to switch it to in terms of the force values. This article also focuses on the design considerations of the sensors since they are delicate, expensive pieces of equipment. Two different methods are used for force resistor sensors first known as thru mode, where there are two polymer layers, each with a semiconductor printed on the thin film; the other is shunt mode which has two thick layers where only one has only one conductive trace to measure the applied forces. Designers in the industry typically try to build around the size of the sensor as they are sometimes bulky. Therefore, multiple designs are required to implement the sensors in the desired way to find the correct parameters [16].

Other sensors are available for different applications to measure different parameters. In this particular article, tire pressure sensors are discussed as they are important for measuring the contact pressure and stress of the tire at various loads and speeds. Some of the sensors are remotely operated, so the sensor can be placed on a rotating drum that the tire is in contact to simulate a rolling road if the wheel is not supported by a vehicle to drive on flat roads. This sensor has multiple applications for evaluating the tire performance at high speeds, identifying failure modes, and measuring the distortion due to inertial forces and cornering performances [17].

This article discusses the physical aspects of the regenerative braking system (RBS). The general design of the system is addressed, using both friction and electrical braking to decelerate the vehicle. It is discussed that the system requires friction braking due to the maximum negative torque that can be used with the motor. As a result, additional braking is needed for scenarios that require more than the motor will allow. Three different strategies are used for RBS high dynamic strategies, and the second and third were different versions of the common strategies. Some examples of the common strategies include serial, eco, and fuzzy logic strategies. The mathematical aspect of the braking forces required is highlighted to address the parameters required in order to achieve a functional regenerative braking system. Calculating this information requires certain values for the car's drive layout, such as its center of gravity height and length of the wheelbase [18].

With electric vehicles becoming more common, new technologies are surfacing to improve their advantages. Some advantages include the large torque output; the motor is small in size and bringing down the overall cost. However, this results in a few drawbacks to the design such as the structure complexity. Due to the increased amount of unsprung mass, the inertia would increase and make it difficult to maintain due to the poor cooling and ventilation. This is important to recognize since the motor cannot reach a certain temperature; otherwise, the permanent magnet will demagnetize and lose all controls. Attempts have been made with different manufacturers for smaller-scale testing where there have been difficulties achieving the target design [19].

Many electric cars implement a four-wheel drive system since the only requirement would be two motors. However, when choosing to use in hub motors to achieve a four-wheel drive layout, many design considerations must be taken into account, which is brought up in this research article. However, for this system, the ideal case cannot be achieved due to its phase inductance and finite inverter voltage. This is then transmitted into a torque ripple pulsates that vary throughout the speed range, which generates vibration and noise throughout the car. A sensor is put in place to suppress the pulsation to reduce this. At lower speeds, the torque ripple is less likely to be corrected by the sensors due to the sensitive noise frequency which it occurs at [20].

A common feature of most modern cars is the stability control system used to minimize the vehicle's rotation speed to prevent the vehicle from achieving too great of a slip angle. Electric vehicles (EVs) with hub motors require direct communication of the vehicle instantaneous dynamics, such as the vehicle yaw and slip angle. These parameters are calculated using linear equations and Newton's law, which can then be rearranged to find the ideal yaw and sideslip angle. Depending on the cars' tire road adhesion, the onboard computer will split the signals to the individual wheels to provide the necessary power at each given moment [21].

When designing an in-hub electric wheel, there are new constraints that designers are faced with as opposed to the regular electric motor. Some of these constraints include the physical, mechanical, and electromechanical. Physical constraints include the size and space available to construct a new design inside the rim of the tire. A mechanical constraint includes the magnitude of the forces influenced by the motor, such as the traction forces. This constraint can vary depending on the road conditions and layout of the car [22].

A brushless DC hub motor is designed, built, tested, and implemented for electrically driven vehicles. Since the stator slots and rotor magnets are orientated to concentrated fractional-pitch winding, the motor is able to achieve a high power density, high efficiency, and no cogging torque. For non-skewed stators, the topology reduces the manufacturing time. This also allows for the motor to be low cost and easily manufactured. The motor can match different speeds and loads where the motor has a high-efficiency region throughout the permitted speed and load range. Using the converter circuit demonstrates the economic and performance feasibility of the design used for two-wheel applications [23].

Using simulation software, energy recovery systems use fuzzy logic controllers to monitor varying loads on the permanent magnet brushless DC motor. Fuzzy logic is a commonly used method to control the dynamic response derived from the performance parameters reflecting the rise, settling, and overshoot. Under its transient conditions, the motor can maintain its rated speed within 0.2 s. In order to output a real value, the system undergoes a conversion called defuzzification. However, the centroid method process is preferred as it takes up less computation time. Through the simulation, the harvesting of the motor improves its overall efficiency. Voltage source inverter is a future implementation to consider, replacing the H-bridge inverter. This would improve torque output and drive, which will reduce current ripples [24].

In 1898, Ferdinand Porsche designed and built his second car with a hybrid-powered system using in-hub electric motors. It was called the Lohner Electric Chaise. This was the world's first front-wheeled drive car and one of the first to have in-hub electric wheel motors. This was a significant point in the history of the hybrid vehicle as it was the front-wheel drive layout that powered the most common hybrid car used to this day, the Toyota Prius [25].

Upon designing an electric vehicle, a few design considerations refer to the power source and propulsion system that has to be decided before pursuing as many other manufacturers have in the past. These include the range of the vehicle due to motor power usage and battery size. How long these components last and how much they will cost? Moreover, the types of energy systems are increasing with more automotive makers making the switch to electric vehicles (EVs) and plugin hybrid electric vehicles (PHEVs). Due to the required rare-earth magnets, permanent magnet (PM) motors are beginning to fall in short supply. As a result, designs such as induction switched reluctance and SynRM motors are going to continue to be used where PM motors are not available. New types of electronic systems will also be used with components such as silicon carbide and GaN-based powered systems to eliminate the requirement for speed and position sensors. Until the power grid has improved, the cost of electricity will go up with the demand due to the increase in the number of electric cars. If these goals are achieved, then the electric vehicles available will better suit the market and the requirements of the future [26].

Electric motors are designed to eliminate the use of unnecessary drivetrain components such as the gearbox. A motor replaces these with interior and exterior rotary structural components. However, the performance of the motor is directly related to the thermal properties during its performance. If the motor got too hot, the demagnetization would decrease its performance and eventually not work altogether. As a result, the design of the magnetization must revolve around a method to cool the system under high load. For specific motors, this is not a problem where there is no contact; however, for brush DC motors, there are contacts directly blocking the airflow of the motor. As a result, that design does not allow for very high efficiency [27].

When designing an in-wheel electric motor, many design restrictions limit the types of motors that can be used. A type of motor commonly used is a dual stator and dual field excitation permanent magnet motor. This is based on the convention of the radial and axial magnetic field motor. This design allows for a higher torque output at lower speeds than the conventional in-wheel motor. Also, the entire speed range does not require the flux weakening control strategy [28].

In order to prevent the vehicle's tires from slipping on the road or artificial road, a traction control system is implemented in the engine management system. However, with the use of in-hub motors, the traction monitoring system has to separate from the motor to apply features such as traction control system, electronic differential system, and electronic stability control. These features are required to ensure the wheels can transfer energy to the road to propel the vehicle. This system can be analyzed using numerical derivations using a microprocessor to calculate the instantaneous solutions [29].

Other advantages can be found by switching the drive motor to in-hub electric wheel. The application of effective chassis control strategies and optimization measurements with motor control through ECU (engine control unit) and crosswire technology will be another advantage of using the electric wheel. The four-wheel drive system on the electric vehicle can perform more effectively with the assistance of the optimized force distribution with the hydraulic braking and motor torque coordination [30].

Few companies have successfully attempted to place the in-hub electric wheels into production due to their exposure to the elements, making it difficult for reliability since it does not have the same protection as the engine bay. Some manufacturers such as Protean have successfully implemented the in-hub motors that have done an extensive job to waterproof and seal the motors as much as possible to keep the life of the motor high. Other drawbacks to the in-hub design are said to be the addition of the unsprung mass. However, after Protean went to Lotus for dynamic simulation and testing, they found that the motor's added mass did not significantly affect the handling, not at least to the untrained driver touch. Much of the noticeable effect was eliminated with the modification of additional suspension damping. Also, each wheel having the ability to power its unsprung mass improved the handling further. This is due to the increase in torque vectoring variability, which means that the torque being sent to each wheel is not constant, which means it is not allowed to change depending on the wheel with the highest traction independently. Unlike a four-wheel drive conventional system, the power is split equally, so there is no torque loss due to slippage. However, this still means that the power is not properly distributed in scenarios with limited surface traction [31].

There are different types of motors used for different applications throughout the industry. One type of motor is the brushed DC motor, which is great for achieving a high amount of torque but is not very efficient from the loss due to friction and requires a lot of maintenance. The motor is also larger than most motors with their large construction and inability to downsize. Another type of motor is the induction motor. This motor has a dependable, durable construction, making it able to work in hostile environments. However, this type of motor requires a specific motor controller, which makes it very expensive. The efficiency is also lower than other motors that are available on the market, such as the permanent magnet brushless DC (PM BLDC) and the switched reluctance motor (SRM). PM BLDC motor is commonly used due to its high efficiency and high power density. High efficiency is due to the use of permanent magnets since they do not require energy to make them magnetic. However, their rarity is also more expensive, and they are trying to phase out as they depend on limited resources. Finally, the best option to select would be the SRM due to its fault-tolerant operation, simple control, and great torque speed available across a constant power range giving it high-speed capabilities. This motor does not feature a magnetic source giving it a low moment of inertia for quick acceleration. The simplistic rotor structure also makes it easy to cool the motor while achieving higher speeds. The SRM, however, does suffer from a torque ripple and acoustic noise, which is common in most in-hub motor designs. Although these faults do not make this type of motor inapplicable to our application, we selected the PM BLDC motor since this is a motor which is widely available and more efficient than the SRM. The efficiency of the motor is essential to our application since we want an accurate testing platform to demonstrate this technology [32].

The permanent magnetic motor is a conventional motor that for this project utilizes three phases. These types of motors have either repulsion or attraction poles. This begins with the current flowing through the three stator windings that allure the PM to the next pole and repeats the cycle to perform the high-speed rotational motion. With each winding charging, the rotor will cause a rotational field. Sensors, such as hall effect sensors, are used to monitor the positioning of the rotor, and it is a change of state with each cycle. However, in this process, the motor produces a ripple in the stator current, which reduces efficiency; with this efficiency loss, the motor will decrease motor power output. Using simulation analysis, however, this can be minimized [33].

Several different options are used in the area of braking systems, such as electrical, dynamic, and mechanical braking. Electrical braking consists of an electrical system using external resistors to provide the braking. This requires the resistance to be very high to prevent large amounts of dissipated heat. Similar to the electrical braking system is dynamic braking, which uses a shunt motor connected as a generator with a field winding parallel with the armature. This is also known as rheostatic braking, which uses external resistance connected across armature terminals. The relationship between the braking torque and resistor values can be represented with a linear graph, which shows that the braking torque will decrease with the increase of the resistor value. Another relation is one between the rpm versus the current, which is another linear relationship. However, higher rpm produces a higher amount of current, which will put more energy back into the system. The traditional method is mechanical braking which utilizes disk rotors and brake pads which is typically used to assist a recovery system since it is not able to harness large amounts of braking torque. However, this means that some of the energy is lost to heat [34].

A common design for electric vehicles is to use only one motor, which is split across two sets of wheels. However, this is not beneficial to the driving dynamic since it means that there is only torque being applied to the powered wheels when approaching a corner. This means the car is not able to travel fast enough where the traction will allow. In a braking scenario, the torque vectoring scenario is improved with the overshoot of the sideslip angle strongly reduced concerning the baseline vehicle. This can be summarized that the torques are proportionally distributed to the vertical load. In another scenario, when traction is present, the feedforward controller frequency domain is required to get the yaw moment required. Similarly, in off-road conditions, using a torque vectoring system increases the power efficiency and performance with a few factors such as rolling resistance and power loss [35].

Important consideration must be made on the car's vehicle dynamics with the addition to in-wheel motors. More specifically, in vehicle dynamics, the parameters affect the sideslip angle and yaw rate. In monitoring the yaw rate values for electric vehicles with four in-wheel motors, yaw controls are placed on the vehicle. These will monitor the change in yaw rate and calculate the correction required and communicate the signal via a CAN (controller area network) bus system. This calculation method uses a fuzzy logic method to determine the correct coefficient requiring the correct yaw rate. However, the amount of sideslip and yaw rate is all limited by the vehicle weight and tire quality, which means a maximum is possible to achieve based on the vehicle parameters. The tire quality can be determined if the control is used enough to determine the tire coefficient since it requires multiple iterations in order to retrieve an average value [36].

To mathematically represent the results to the driver input, functions can be made in order to best find the resulting correction for the yaw rate and sideslip angles. A sine curve can approximate the wheel steering angle under a certain driving condition. While maintaining a constant velocity, the single wheel steering angle can be set to vary depending on the state of the multiple sine curves. The response of the actual yaw rate with and without control to the nominal yaw rate under single-lane driving conditions coincides with the actual side slip angle at the mass center with and without control [37].

Introducing the switch is helpful for larger electric vehicles only when there is enough space. However, for smaller electric vehicles, packaging the energy systems and motor in the vehicle can be challenging. Therefore, the ideal alternative would be to free space in the car by using an in-wheel electric motor. This allows for the vehicle to have a compact structure, large usage of interior space, and lower vehicle center of gravity, which means the car can be more stable. Lowering the center of gravity will improve stability. The added unsprung mass will have an increased amplitude in vibrations and negatively affect the handling. In view of these advantages, more manufacturers are attracted to electric wheel technology while overcoming the challenges [38].

There are many advantages to the vehicle's powertrain to in-wheel motor design; however, making this decision has some drawbacks that affect the vehicle's quality. In most in-wheel motor designs, vibrations are generated due to an interaction between the torque ripple and the rotor field, which reaches a high level of harmonics. This can be seen with the concentrated winding of the stator field harmonics. Although this is present with all types of motors, the switched reluctance motor is generally preferred for the distributed winding type, whereas with permanent magnet synchronous machines, a concentrated winding has limitations due to typically high stator field harmonics [39].

It is important to understand the kinetic energy produced by the electric wheel. Also, it is essential to analyze the behavior of the electric motor under the motor battery braking efficiency. This can be achieved using the QS motor, which produces 8000W that when placed with the motor and battery, the motor acts as a generator. To increase the braking torque required to run the generator, additional power is provided to the circuit to slow the wheel further and recover the energy to the electric energy system [40].

Different companies explore different areas of regenerative braking to improve electric vehicle's power usage. One of the areas in the regenerative braking field is the use of E-REV bus, a type of coast that utilizes the kinetic energy in the vehicle during the acceleration and uses the stored energy without the use of additional energy and the coasting of the vehicle. This can be compared with the regenerative braking system to the consistent driving style. Studies have shown that when coasting with regenerative braking, this shows an increase in 3.5% in fuel economy. When comparing the use of coasting to traditional braking, the fuel economy can improve by 39.7%. Although it isn't a significant factor, the regenerative braking does have a role in improving the fuel economy [41].

The motor simulation is used to compare the ideal scenario versus the actual testing. There can be a better understanding of the parameters evaluated to test the performance properly. For permanent magnets, brushless motors are evaluated as being better for a high-performance scenario while under the characteristics of low-speed range, high torque and power density. The mathematical model, which represents this simulation, has commonly consisted of four blocks, a BLDC block, an inverter, a controller, and a subsystem, much like the required physical wiring components. In the model, PI controllers are used to monitor the motor's speed, adjusting the duty cycle of the devices it simulates in using. The phase curves produced by this simulation method are sinusoidal with a phase difference of 120 degrees. Optimum performance parameters such as torque and rotational speed can be found. However, a load must be applied to find an accurate value for the torque. This will produce a precise waveform of back EMF (electromagnetic force), which can then be used to determine the torque values [42].

In order to replicate the performance of the motor, a simulation must be made using closed-loop control of a permanent magnet brushless DC motor. This simulation performance pattern is in the form of a PID controller, which produces a waveform profile that allows the torque and time, speed and time, current and time, or voltage and time to be obtained. Different results can be given with the simulation depending on the conditions with a specific load or with no load to understand the results better. Using a PID controller will enable the accurate calculation of error by comparing the referenced data with the real data. This can be tuned to better match the ideal data by changing the proportional, integral, and derivative terms. This will accurately represent how the motor should perform [43].

The research on energy storage systems will enhance the overall efficiency of electric wheel deployments. However, there are situations where energy is lost, such as under braking. Since generating larger batteries is complex, the alternative is utilizing fuzzy control strategies to enable the motor to put energy back into the battery, which will restore some of the lost range. Specific implementation includes the usage of hydraulic systems while generating electricity. This is called a parallel braking system, where the regenerative torque from the motor acts on the driving axle. A layout used is the LF620 layout, which utilizes a motor, pump, accumulator, and pressure regulating valve. Under braking, the VCU (vehicle control unit) finds the proper deceleration and target brake force. The pedal is then applied using a vacuum booster, where the rear wheels supply the pressure from the master cylinder through a proportional valve [44].

While the road transport emissions are responsible for two-thirds of the world's CO_2 transportation emissions, it has been steadily growing in developing economies around the world, with the number of vehicles having increased by 56% from 1990 to 2007. With this being only 17 years, there is a significant dependency on personal transportation as the number of vehicles being placed on the road will increase over the next few decades. This means that there is a higher demand for electrically powered vehicles in order to reduce the average amount of CO_2 in the air to decrease the rising planet temperature. Other problems that need to be addressed include the selection of energy sources. As of the year, 2000, China has been the leader in electrical infrastructure, generating nearly one-quarter of the electrical generating capacity commission through coal-burning facilities. While China is still a young energy producer, they account for 37% of the world's emissions, which shows that the production method needs to change if there is to be a decrease in harmful emissions [45].

12.4 EV with Electric Wheel

The electric wheel (EW) is integrated within EV so that battery can be connected with all left and right electric wheels. The generator is used for regenerative braking. Hybrid electric wheel and the regular wheel are configured for both left and right sides, as shown in Fig. 12.6. The design and configuration of the hybrid electric wheel will be optimized to achieve the highest performance.

The analysis will include the dynamic behavior of the vehicle and wheels with different road friction, slope, turning angle, vehicle load, and weather conditions. Dynamic system modeling with the governing equations is shown in [1]. Improved design can be achieved with coordinated motion with the power train controllers.

12.5 Summary

Electric wheel (EW) or in-wheel electric motor can offer improved performance of EVs with enhanced stability, energy efficiency, and reliability of EV. EW will enhance the regenerative braking system where it is connected to each EW. The review of the regenerative braking systems is explained with related concepts and technologies. The analysis of electric wheel design and related components is discussed to provide good understanding of functions and techniques involved in the development of the electric wheel. The configuration of electric wheels with conventional wheels in EV can be defined in hybrid ways to maximize the overall performance of EV. The concept of EW can be retrofitted in existing internal combustion vehicles and EV. In addition, EW could be installed in any physical system to offer the required mobility based on user requirements. The charging of EV with EW can be implemented with considerations of energy loads and specifications of EWs.

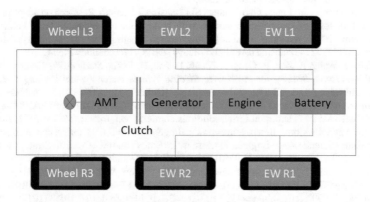

Fig. 12.6 Hybrid electric wheel configuration within EV

Acknowledgment The author would like to thank Myles Ginty, Seyawash Nairow, Philip Anisimov, and Ahmed Abdelmaksoud who supported this research. Also, the author thanks other members from the Smart Energy Systems Laboratory (SESL) at Ontario Tech University.

References

1. Z. Shuai, C. Li, J. Gai, Z. Han, G. Zeng, G. Zhou, Coordinated motion and powertrain control of a series-parallel hybrid 8 × 8 vehicle with electric wheels. Mech. Syst. Signal Process. **120**, 560–583 (2019)
2. W. Zhao, Z. Yang, C. Wang, Multidisciplinary hybrid hierarchical collaborative optimization of electric wheel vehicle chassis integrated system based on driver's feel. Struct. Multidiscipl. Optim. **57**(3), 1129–1147 (2017)
3. Cars, https://www.cars.com/auto-repair/glossary/differential/. Accessed 6 Jan 2022
4. D. Lippert, J. Spektor, Rolling resistance and industrial wheels, Hamilton Caster [Online]. Available http://www.mhi.org/media/members/14220/130101690137732025.pdf
5. Rimac, Rimac all-wheel torque vectoring. Accessed 6 Jan 2021 [Online]. https://www.rimac-automobili.com/en/press/news/rimac-all-wheel-torque-vectoring/
6. G. Solberg, The magic of tesla roadster regenerative braking, Tesla. Accessed 6 Jan 2021 [Online]. https://www.tesla.com/blog/magic-tesla-roadster-regenerative-braking
7. L. Wang, D.-f. Li, H.-x. Xu, Z.-p. Fan, W.-b. Dou, X.-l. Yu, Research on a pneumatic hybrid engine with regenerative braking and compressed-air-assisted cranking, Sage J, June 4, 2015. Available [Online] https://doi-org.uproxy.library.dc-uoit.ca/10.1177%2F0954407015586706
8. R.R.S. Bravo, V.J. De Negri, A.A.M. Oliveira, Design and analysis of a parallel hydraulic – pneumatic regenerative braking system for heavy-duty hybrid vehicles, ScienceDirect, 12 May 2018. Available [Online]. https://www-sciencedirect-com.uproxy.library.dc-uoit.ca/science/article/pii/S0306261918306688#f0090
9. J. Bian, B. Qiu, Effect of road gradient on regenerative braking energy in a pure electric vehicle. Sage Journals, November 14, 2017. Available [Online] https://journals-sagepub-com.uproxy.library.dc-uoit.ca/doi/full/10.1177/0954407017735020?utm_source=summon&utm_medium=discovery-provider
10. L. Li, X. Li, X. Wang, Y. Liu, J. Song, X. Ran, Transient switching control strategy from regenerative braking to anti-lock braking with a semi-brake-by-wire system. Taylor and Francis online, 8 Jan 2016. Available.[Online]. https://www-tandfonline-com.uproxy.library.dc-uoit.ca/doi/full/10.1080/00423114.2015.1129059
11. Sun Chengwei; Chu Liang; Guo Jianhua; Shi Dapai ; Li Tianjiao, Research on adaptive cruise control strategy of pure electric vehicle with braking energy recovery, Sage Publications Ltd, Nov 2017. Available [Online] https://search-proquest-com.uproxy.library.dc-uoit.ca/docview/1977721438/abstract/A65F2B57828947DBPQ/1?accountid=14694
12. Y. Sun, Y. Wang, R. Zhu, R. Geng, J. Zhang, D. Fan, H. Wang, *Study on the Control Strategy of Regenerative Braking for the Hybrid Electric Vehicle under Typical Braking Condition*, IOP Publishing Ltd, 13 Dec 2018. Available [Online]. http://fr7cx7ua3s.search.serialssolutions.com/?ctx_ver=Z39.88-2004&ctx_enc=info%3Aofi%2Fenc%3AUTF-8&rfr_id=info%3Asid%2Fsummon.serialssolutions.com&rft_val_fmt=info%3Aofi%2Ffmt%3Akev%3Amtx%3Ajournal&rft.genre=proceeding&rft.title=IOP+Conference+Series%3A+Materials+Science+and+Engineering&rft.atitle=Study+on+the+Control+Strategy+of+Regenerative+Braking+for+the+Hybrid+Electric+Vehicle+under+Typical+Braking+Condition&rft.au=Sun%2C+Yuantao&rft.au=Wang%2C+Yunlong&rft.au=Zhu%2C+Rongfu&rft.au=Geng%2C+Ruiguang&rft.date=2018-12-13&rft.issn=1757-8981&rft.eissn=1757-899X&rft.volume=452&rft.issue=3&rft_id=info:doi/10.1088%2F1757-899X%2F452%2F3%2F032092&rft.externalDBID=n%2Fa&rft.externalDocID=626927275&paramdict=en-US

13. Chen Lv; Junzhi Zhang; Yutong Li; Ye Yuan, Mechanism analysis and evaluation methodology of regenerative braking contribution to energy efficiency improvement of electrified vehicles, Science Direct, 16 January 2015. Available. [Online]. https://www-sciencedirect-com.uproxy.library.dc-uoit.ca/science/article/pii/S0196890414011418

14. M. Santora, Bearing friction basics: A primer, Bearing Tips, Accessed 6-Jan-2022. [Online] https://www.bearingtips.com/bearing-friction-basics-primer/

15. Engineering ToolBox, Rolling Resistance., Accessed 6-Jan-2022, [Online]. https://www.engineeringtoolbox.com/rolling-friction-resistance-d_1303.html (2008)

16. Machine Design, THE DIFFERENCE BETWEEN Force Measurement Techniques, Tekscan, March 2018. [PDF] Accessed 6-Jan-2022: https://www.tekscan.com/thank-you/download-now-difference-between-force-measurement-techniques

17. Tekscan, High-Speed TireScan™ System Capture & Analyze Tire Behavior at High Speeds (2018) [Online]. Accessed 6 Jan 2022 https://www.tekscan.com/products-solutions/systems/high-speed-tirescan

18. B. Xiao, H. Lu, H. Wang, J. Ruan, N. Zhang, Enhanced Regenerative Braking Strategies for Electric Vehicles: Dynamic Performance and Potential Analysis, energies, MDPI, 15 November 2017. [Online]. Available https://res.mdpi.com/energies/energies-10-01875/article_deploy/energies-10-01875-v2.pdf?filename=&attachment=1

19. L. Jian, Research Status and Development Prospect of Electric Vehicles Based on Hub Motor, IEEE, 31 December 2018 [Online]. Available: https://ieeexplore-ieee-org.uproxy.library.dc-uoit.ca/document/8592598/references#references

20. D. Lu, J. Li, M. Ouyang, J. Gu, Research on hub motor control of four-wheel-drive electric vehicle, IEEE, 13 October 2011 [Online]. Available: https://ieeexplore-ieee-org.uproxy.library.dc-uoit.ca/document/6043150

21. N-N. Zhang, H-L. Li, Y-Y. Li, Stability control of hub motor driven pure electric vehicle, IEEE, 23 October 2017 [Online]. Available https://ieeexplore-ieee-org.uproxy.library.dc-uoit.ca/document/8079866

22. K. Cakir, A. Sabanovic, In-wheel motor design for electric vehicles, IEEE Explore, 30 May 2006. [Online]. Available https://ieeexplore-ieee-org.uproxy.library.dc-uoit.ca/document/1631730

23. S. Ekram, D. Mahajan, M. Fazil, V. Patwardhan, N. Ravi, Design optimization of brushless permanent magnet hub motor drive using FEA, IEEE, 26 December 2007 [Online]. Available https://ieeexplore-ieee-org.uproxy.library.dc-uoit.ca/document/4412273

24. J. Joy, S. Ushakumari, Energy harvestation and effective utilization by regenerative braking and motoring modes of operation in a PMBLDC drive system using FLC, IEEE, 19 October 2017. [Online]. Available. https://ieeexplore-ieee-org.uproxy.library.dc-uoit.ca/document/8069991

25. Hybrid Cars, History of hybrid vehicles, March 27, 2006 [Online]. Accessed 6 Jan 2022. https://web.archive.org/web/20090904154040/http://www.hybridcars.com/history/history-of-hybrid-vehicles.html

26. Kaushik Rajashekara, Present Status and Future Trends in Electric Vehicle Propulsion Technologies, IEEE, 23 April 2013 [Online]. Available https://ieeexplore-ieee-org.uproxy.library.dc-uoit.ca/document/6507304

27. K. Roumani, B. Schmuelling, Electromagnetic design of central motor and in-wheel hub motor according to their voltage range, IEEE, 02 February 2017 [Online]. Available https://ieeexplore-ieee-org.uproxy.library.dc-uoit.ca/document/7837252

28. P. Gao, Y. Gu X. Wang, The design of a permanent magnet in-wheel motor with dual-stator and dual-field-excitation used in electric vehicles. Energies, 12 February 2018. [Online] Available https://www.mdpi.com/1996-1073/11/2/424/htm

29. J-S. Hu, X-C. Lin, D. Yin, F.-R. Hu, Dynamic motion stabilisation for front-wheel-drive in-wheel motor electric vehicles. Advances in Mechanical Engineering; New York, Dec 2015 [Online]. Available https://search-proquest-com.uproxy.library.dc-uoit.ca/docview/1771393875/citation/4005D8E7A3F4743PQ/1?accountid=14694

30. H. Zhang, W. Zhao, Decoupling control of steering and driving system for an in-wheel-motor-drive electric vehicle, College of Energy and Power Engineering, Nanjing University of Aeronautics and Astronautics, No. 29, Yudao Street, Nanjing 210016, China, 23 September 2017. [Online]. Available. https://www-sciencedirect-com.uproxy.library.dc-uoit.ca/science/article/pii/S0888327017304703

31. A. Whitehead, C. Hilton, Protean Electric's In-Wheel Motors Could Make EVs More Efficient, IEEE Spectrum, 15:00 GMT, 26 Jun 2018 [Online]. Available https://spectrum.ieee.org/transportation/advanced-cars/protean-electrics-inwheel-motors-could-make-evs-more-efficient

32. X.D. Xue, K.W.E. Cheng, N.C. Cheung, Selection of Electric Motor Drives for Electric Vehicles, IEEE, 10 April 2009 [Online]. Available https://ieeexplore.ieee.org/document/4813059

33. S.S. Gotkhinde, P.R. Jadhav, Harmonic analysis of PMBLDC motor drive, IEEE, 23 November 2017. [Online]. Available. https://ieeexplore-ieee-org.uproxy.library.dc-uoit.ca/document/8117966

34. 1 in. Standard Duty Pillow Block Bearing Assembly. Princess Auto. Accessed 6 Jan 2022. https://www.princessauto.com/en/detail/1-in-standard-duty-pillow-block-bearing-assembly/A-p3870052e

35. L. De Novellis, A. Sorniotti, P. Gruber, L. Shead, V. Ivanov, K. Hoepping, Torque Vectoring for Electric Vehicles with Individually Controlled Motors: State-of-the-Art and Future Developments, Scopus, World Electric Vehicle Journal, 2012. Available [Online]. https://pdfs.semanticscholar.org/3daf/4e8af7e798dc1b82cd4f06b246b27d1378fc.pdf

36. Rufei Hou; Li Zhai; Tianmin Sun Steering stability control for a four hub-motor independent-drive electric vehicle with varying adhesion coefficient, Energies, 14 September 2018. Available. [Online]. https://www.mdpi.com/1996-1073/11/9/2438

37. B. Jin, C. Sun, X. Zhang, Research on lateral stability of four hub motor-in-wheels drive electric vehicle. Int. J. Smart Sens. Intell. Syst. 8(3), September 1, 2015. Available [Online]. http://fr7cx7ua3s.search.serialssolutions.com/?ctx_ver=Z39.88-2004&ctx_enc=info%3Aofi%2Fenc%3AUTF-8&rfr_id=info%3Asid%2Fsummon.serialssolutions.com&rft_val_fmt=info%3Aofi%2Ffmt%3Akev%3Amtx%3Ajournal&rft.genre=article&rft.atitle=Research+on+lateral+stability+of+four+hub-motor-in-wheels+drive+electric+vehicle&rft.jtitle=International+Journal+on+Smart+Sensing+and+Intelligent+Systems&rft.au=Jin%2C+Biao&rft.au=Sun%2C+Chuanyang&rft.au=Zhang%2C+Xin&rft.au=Zhang%2C+Xin&rft.date=2015&rft.eissn=1178-5608&rft.volume=8&rft.issue=3&rft.spage=1855&rft.epage=1875&rft_id=info:doi/10.21307%2Fijssis-2017-833&rft.externalDBID=n%2Fa&rft.externalDocID=608969213&paramdict=en-US

38. Y. Sun, M. Li, C. Liao, Analysis of wheel hub motor drive application in electric vehicles, EDP Sciences, 2016. Available [Online] https://www.matec-conferences.org/articles/matecconf/abs/2017/14/matecconf_gcmm2017_01004/matecconf_gcmm2017_01004.html

39. G. Freitag, M. Klöpzig, K. Schleicher, M. Wilke, M. Schramm, High-performance and highly efficient electric wheel hub drive in automotive design, IEEE, 23 December 2013. Available [Online]. https://ieeexplore.ieee.org/document/6689736/authors#authors

40. X. Huang, J. Wang, Model predictive regenerative braking control for lightweight electric vehicles with in-wheel motors, Sage J, April 25, 2012. Available [Online].

41. J. Choi, J. Jeong, Y.-i. Park, S.W. Cha, Analysis of regenerative braking effect to improve fuel economy for E-REV bus based on simulation, 2015. Available [Online]. https://www.mdpi.com/2032-6653/7/3/366

42. S. Dudhe Shivraj, G. Thosar Archana, Mathematical Modelling And Simulation Of Three Phase BLDC Motor Using Matlab/Simulink, I A E T Publishing Company, Nov 2014. Available [Online]. http://search.proquest.com.uproxy.library.dc-uoit.ca/docview/1625941907?accountid=14694

43. K. Avanish, Close Loop Speed Controller for Brushless DC Motor for Hybrid Electric Vehicles, Springer Singapore, 01/01/2019. Available [Online]. http://fr7cx7ua3s.search.serialssolutions.com/?ctx_ver=Z39.88-2004&ctx_enc=info%3Aofi%2Fenc%3AUTF-8&rfr_id=info%3Asid%2Fsummon.serialssolutions.com&rft_val_fmt=info%3Aofi%2Ffmt%3

Akev%3Amtx%3Abook&rft.genre=proceeding&rft.title=Lecture+Notes+in+Electrical
+Engineering&rft.atitle=Close+loop+speed+controller+for+brushless+DC+motor+for+
hybrid+electric+vehicles&rft.au=Kumar%2C+Avanish&rft.au=Thakura%2C+P.R&rft.
date=2019-01-01&rft.isbn=9789811307751&rft.issn=1876-1100&rft.eissn=1876-1119&rft.
volume=511&rft.spage=255&rft.epage=268&rft_id=info:doi/10.1007%2F978-98
1-13-0776-8_24&rft.externalDBID=n%2Fa&rft.externalDocID=623568246&param
dict=en-US

44. Guoqing Xu; Weimin Li; Kun Xu; Zhibin Song, An Intelligent Regenerative Braking Strategy
for Electric Vehicles, MDPI Energies, 22 September 2011. Available [Online]. https://www.
mdpi.com/1996-1073/4/9/1461/htm

45. S.J. Davis, K. Caldeira, H.D. Matthews, Future CO_2 Emissions and Climate Change from
Existing Energy Infrastructure, American Association for the Advancement of Science,
10 September 2010. Available [Online]. https://www-jstor-org.uproxy.library.dc-uoit.ca/
stable/41075805

46. Inside EVs, Accessed 6 Jan 2022. https://insideevs.com/news/320023/protean-teams-up-with-
volkswagen-to-develop-production-in-wheel-electric-motor/

Chapter 13
Fast-Charging Infrastructure Planning

Hossam A. Gabbar

13.1 Introduction

Chargers for transportation electrification could serve different transportation categories such as a vehicle, bus, truck, boat, or rail. Chargers could be off-route (e.g., at bus depots) or on-route (e.g., at bus terminals) for mass transit. This chapter will focus on planning fast-charging stations, which could be at the depot with multiple purposes. Fast-charging stations could offer a number of implementation options depending on charging requirements and the potential of deployment of hybrid energy storage platform of hybrid flywheel technology, ultracapacitor, and battery systems. The planning will consider both functions, fast charging for transportation electrification and the utilization of hybrid energy storage platform to offer a fast response of charge/discharge to balance energy loads in substations and industrial applications. The transition to electric buses is studied, and the planning problem is discussed [1]. Bus electrifications will impact community and marketplace, as discussed in [2]. The charging infrastructure is evaluated as per electric bus fleet size and mixture with other transportation networks, as discussed in [3]. There are direct impacts on the grid when interfacing bus operation and charging as part of demand response and load management, which is also linked to tariff [4]. Studies analyzed the expansion of public transportation networks and their interactions with the grid [5–6]. The analysis of fast-charging deployment to support electric buses showed the improved economic impacts considering electricity demand charges [7]. Case studies are analyzed in different places around the world, such as the European case

---

This chapter is contributed by Hossam A. Gabbar.

H. A. Gabbar (✉)
Energy Systems and Nuclear Science, University of Ontario Institute of Technology,
Oshawa, ON, Canada
e-mail: Hossam.gabbar@uoit.ca

discussed in [8], in Canada as discussed in [9], in London [10], and in Oporto [11]. Design requirements and features of fast-charging stations are developing with the incorporation of smart features [12], while considering energy storage requirements [13], and optimization [14]. The charging station planning is also analyzed given battery swapping strategies [15]. The design of fast-charging station is improved by introducing flywheel energy storage systems [16]. The proposed flywheel-based fast-charging station is adopted to plan the deployment of fast-charging infrastructures, as presented in the following sections.

13.2   Charging Load Analysis

This section will discuss charging load analysis for different transportation systems, including electric vehicle, electric bus, electric truck, electric rail, marine electrification, charging load for substations, charging loads for industrial facility, and an integrated view of charging load in a given region.

13.3   Load Profiles of EVs

In order to properly plan fast-charging infrastructures, it is essential to understand mobility demands and routes of EV, which include the following parameters for trip number "i":

- $EV_{spi}(i)$: Start point location index (city center, industrial area, airport, etc.)
- $EV_{sp}(i)$: Start point location
- $EV_{epi}(i)$: End point location index (city center, industrial area, airport, etc.)
- $EV_{ep}(i)$: End point location
- $EV_{td}(i)$: Trip distance
- $EV_{mi}(i)$: Mobility reason index (business, study, health, tourism, recreation, family, shopping, etc.)
- $EV_{dt}(i)$: Departure time of the day
- $EV_{at}(i)$: Arrival time of the day
- $EV_{vt}(i)$: Vehicle type, model
- $EV_{w}(i)$: Weight of EV
- $EV_{np}(i)$: Number of passengers in vehicle
- $EV_{pi}(i)$: Passengers category index
- $EV_{vp}(i)$: Vehicle priority index
- $EV_{ti}(i)$: Traffic congestion index (reflected to number of braking during trip)
- $EV_{ri}(i)$: Road condition index (clear, snow buildup, construction, turns, speed, etc.)
- $EV_{wi}(i)$: Weather condition index (clear, storm, wind, rain, snow)
- $EV_{l}(i)$: EV load (which will be estimated based on the above parameters)

The analysis of EV charging load is mapped to the selected region, as shown in Fig. 13.1. The start and end points are modeled over given time period for each EV and in view of the defined route.

The total charging loads at each point in the selected region will be used to estimate and optimize the location and size of charging stations for EVs in each region based on EV trips. Figure 13.2 shows an integrated view of possible charging load profiles for different EVs in the given region over the time period.

At each point, the total charging load profiles are estimated based on different trips and routes for the given region and date range. Optimization algorithm based on AI (artificial intelligence) could be used to identify the most suitable location and size of charging stations.

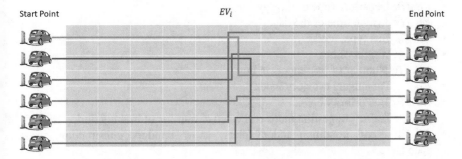

Fig. 13.1 EV load planning

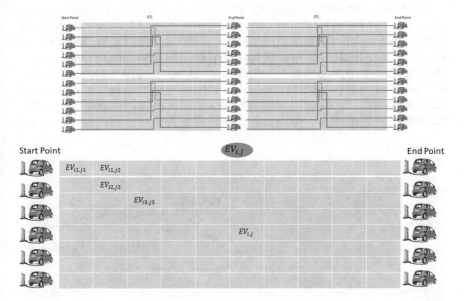

Fig. 13.2 Integrated EV charging load profiles for the given region

13.4   Load Profiles of e-Buses

Similar to EV, the planning of electric buses will be performed based on bus routes, which include the following parameters for trip number "i":

- $EB_{spi}(i)$: Start point location index (city center, industrial area, airport, etc.)
- $EB_{sp}(i)$: Start point location
- $EB_{epi}(i)$: End point location index (city center, industrial area, airport, etc.)
- $EB_{ep}(i)$: End point location
- $EB_{td}(i)$: Trip distance
- $EB_{mi}(i)$: Mobility reason index (business, study, health, tourism, recreation, family, shopping, etc.)
- $EB_{dt}(i)$: Departure time of the day
- $EB_{at}(i)$: Arrival time of the day
- $EB_{vt}(i)$: Bus type, model
- $EB_w(i)$: Weight of bus
- $EB_{np}(i)$: Number of passengers in bus
- $EB_{pi}(i)$: Passengers category index
- $EB_{vp}(i)$: Bus priority index
- $EB_{ti}(i)$: Traffic congestion index (reflected to number of braking during trip)
- $EB_{ri}(i)$: Road condition index (clear, snow buildup, construction, turns, speed, etc.)
- $EB_{wi}(i)$: Weather condition index (clear, storm, wind, rain, snow)
- $EB_l(i)$: e-Bus load (which will be estimated based on the above parameters)

The analysis of charging loads for electric buses in the selected region is estimated based on routes for different buses, as shown in Fig. 13.3. The total bus loads for the given region will be analyzed based on different bus routes, which will be used to estimate total charging loads for each point. The most suitable location of bus charging stations, type, size, and configuration will be evaluated and optimized using AI algorithms. It is possible to consider V2V (vehicle-to-vehicle) charging models and incorporate in the charging planning process to optimize cost and time and other performance measures key performance indicators (KPIs) for individual buses and

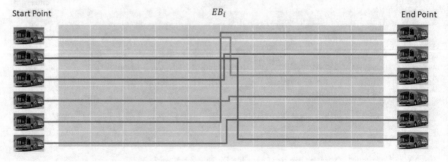

Fig. 13.3 e-Bus load planning

overall parties, including utility companies, transit companies, end users, and transportation technology providers.

13.5   Load Profiles of e-Trucks

Similar to EV, the planning of electric trucks will be performed based on truck routes and required distribution, which includes the following parameters for trip number "i":

- $ET_{spi}(i)$: Start point location index (city center, industrial area, airport, etc.)
- $ET_{sp}(i)$: Start point location
- $ET_{epi}(i)$: End point location index (city center, industrial area, airport, etc.)
- $ET_{ep}(i)$: End point location
- $ET_{td}(i)$: Trip distance
- $ET_{g}(i)$: Goods index (food, electronics, stationery, clothes, etc.)
- $ET_{dt}(i)$: Departure time of the day
- $ET_{at}(i)$: Arrival time of the day
- $ET_{vt}(i)$: Truck type, model
- $ET_{w}(i)$: Weight of truck
- $ET_{vp}(i)$: Truck priority index
- $ET_{ti}(i)$: Traffic congestion index (reflected to number of braking during trip)
- $ET_{ri}(i)$: Road condition index (clear, snow buildup, construction, turns, speed, etc.)
- $ET_{wi,\,i}$: Weather condition index (clear, storm, wind, rain, snow)
- $ET_{l,\,i}$: e-Truck load (will be estimated based on the above parameters)

The analysis of charging loads for electric trucks in the selected region is estimated based on routes for different trucks. Figure 13.4 shows starting and ending points of a number of electric trucks with different routes. The total electric truck loads for the given region will be analyzed based on different truck routes, which will be used to estimate total charging loads for each point in the overall region. The most suitable location of truck charging stations, type, size, and configuration will

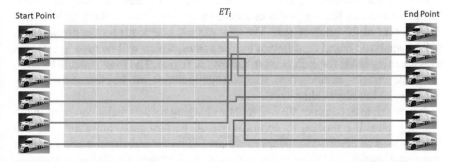

Fig. 13.4   e-Truck load planning

be evaluated and optimized using an intelligent algorithm based on AI. It is possible to consider V2V or mobile charging models and incorporate in the charging planning process to optimize overall performance for individual trucks and associated parties, including utility companies, distribution companies, stores, and transportation technology providers. Performance measures include cost and time, as well as other performance measures (KPIs).

13.6   Load Profiles of Electric Marine

Similar to EV, the planning of electric boats and other marine systems will be performed based on boat routes in offshore, which include the following parameters for the trip number "i":

- $EM_{spi}(i)$: Start point location index (city center, industrial area, airport, etc.)
- $EM_{sp}(i)$: Start point location (in the shore side)
- $EM_{td}(i)$: Trip distance
- $EM_{mi}(i)$: Mobility reason index (business, shipping, fishing, tourism, recreation, family, etc.)
- $EM_{dt}(i)$: Departure time of the day
- $EM_{at}(i)$: Arrival time of the day
- $EM_{vt}(i)$: Boat type, model
- $EM_{w}(i)$: Weight of boat
- $EM_{np}(i)$: Number of passengers in boat
- $EM_{pi}(i)$: Passengers category index
- $EM_{vp}(i)$: Boat priority index
- $EM_{ti}(i)$: Traffic congestion index (as number of boats or activities in offshore during trip)
- $EM_{ri}(i)$: Water condition index (clear, snow, wave, storm, etc.)
- $EM_{wi}(i)$: Weather condition index (clear, storm, wind, rain, snow)
- $EM_{l}(i)$: e-Boat load (which will be estimated based on the above parameters)

The analysis of total loads for charging electric boats and other electric marine systems in the selected region is estimated based on routes for different trucks. Figure 13.5 shows starting points of a number of electric boats along with the different routes in the offshore side. The total electric boat loads for the given region will be analyzed based on different marine routes, which will be used to estimate total charging loads for each point in the overall region. The most suitable location of marine charging stations, type, size, and configuration will be evaluated and optimized using intelligent algorithms.

It is possible to consider V2V or mobile charging models and incorporate in the charging planning process to optimize overall performance, including cost and time, as well as other performance measures (KPIs) for individual e-boats and the associated parties, including utility company, boat company, shipping companies, city plan, and marine transportation technology providers.

Charge Point EM_i

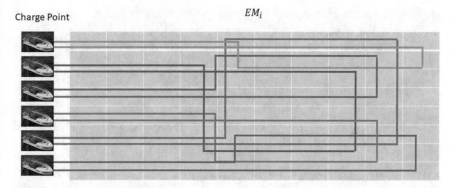

Fig. 13.5 Electric marine load planning

13.7   Load Profiles for Power Substations

This section explains the planning and implementation of high-performance and resilient hybrid energy storage with fast-charging station to support and balance energy demand and supply differentials at the power substation. The target fast-charging station will support transportation electrification infrastructures with potential charging station near the substation to support different transportation routes. In addition, the fast-charging station will support surrounding residential and industrial facilities.

In order to plan the fast-charging station as integrated with power substations, detailed designs and key parameters from each substation and grid and transmission lines will be analyzed. The following parameters for each substation "i" will be considered:

- $SS_{li}(i)$: Station location index (city center, industrial area, airport, etc.)
- $SS_l(i)$: Station location
- $SS_{ul}(i)$: Maximum (upper) load
- $SS_{ll}(i)$: Minimum (lower) load
- $SS_{sc}(i)$: Number of served clients
- $SS_{ld}(i)$: Load differential profile
- $SS_{hc}(i)$: Station health condition
- $SS_{wi}(i)$: Weather condition index (clear, storm, wind, rain, snow)
- $SS_l(i)$: Substation charging demand (which will be estimated based on the above parameters)

The distribution of charging stations in the selected region is shown in Fig. 13.6 where fast-charging station is planned beside each substation. The planning of the fast-charging station will include the design of several energy storage models using battery systems and flywheel technologies based on design requirements, load profiles, and other system specifications. Configuration parameters will be defined and optimized based on the most suitable operational scenarios. Comprehensive economic and technical performance indicators will be developed and evaluated for the

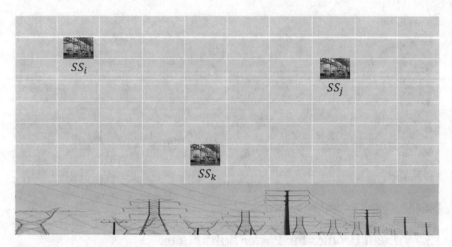

Fig. 13.6 Charging stations associated with power substations

design and operation of the proposed integrated system. Risks and design fault-tolerant and self-healing approaches will be assessed to ensure resilient energy storage systems. The overall performance, including power flow, costs, quality, resiliency, and risks, will be evaluated and optimized with environmental consideration for the potential design alternatives and control scenarios. Installation and testing of the integrated fast charging and substation will be performed to ensure optimum overall performance.

13.8   Load Profiles for Industrial Facilities

Charging stations are required for industrial facilities (including distribution centers) to support electric machines, peak load shaving, and backup power supply during emergencies. Also, the fast-charging station will be used to charge electric batteries used in the industrial facility to support daily operation. Figure 13.7 shows possible locations of industrial facilities in the given region where charging stations are integrated with each industrial facility. This will also support transportation electrification with stations near these industrial facilities.

In order to plan the fast-charging station as integrated with industrial facilities, designs and main parameters from each industrial facility will be analyzed. The following parameters for each industrial facility "i" will be considered:

- $IF_{li}(i)$: Industrial facility location index (city center, industrial area, airport, etc.)
- $IF_l(i)$: Industrial facility location
- $IF_s(i)$: Industrial facility size
- $IF_{ul}(i)$: Maximum (upper) load
- $IF_{ll}(i)$: Minimum (lower) load
- $IF_{ic}(i)$: Industrial category (distribution center, factory, etc.)

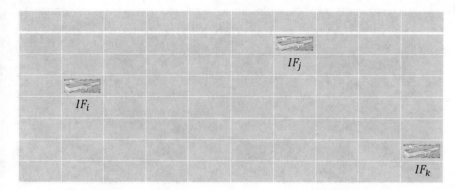

Fig. 13.7 Charging load for industrial facilities

- *IF$_{ld}$(i)*: Load differential profile, which is supported by fast-charging station
- *IF$_{es}$(i)*: Industrial facility total battery storage required for electric machines
- *IF$_{hc}$(i)*: Industrial facility health condition
- *IF$_{wi}$(i)*: Weather condition index (clear, storm, wind, rain, snow)
- *IF$_l$(i)*: Industrial facility charging demand (which will be estimated based on the above parameters)

The design and operation parameters of the planned fast-charging station will be optimized in view of industrial facility parameters, grid condition, and transportation electrification requirements. The overall performance will be optimized, including cost and time and other performance measures (KPIs) for individual industrial facilities.

13.9   Integrated Load Profiles

This section discussed the integrated planning of total charging demands for all possible options in the selected region, which include electric vehicle, electric bus, electric truck, electric boat, substation, and industrial facility. Example model is shown in Fig. 13.8. The models described in the above sections will be used to integrate and used to analyze the augmented charging demands. The location, size, and type of charging stations will be optimized in view of individual performance measures and the overall performance measures.

13.10   Development of Fast-Charging Station for Industrial Facilities and e-Trucks

Commercial and industrial facilities require a stable and clean power supply to support production activities during normal and abnormal operations, critical systems and activities, and power outages. Some industries utilize battery storage

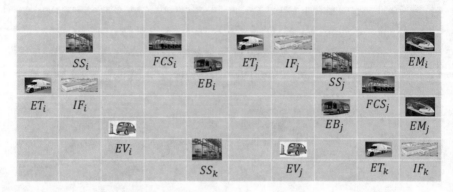

Fig. 13.8 Integrated planning of charging loads

technologies as part of their production systems and equipment and in maintenance activities. There are limitations in current charging stations that take a longer time with reduced performance, higher lifecycle costs, and inability to provide resilient energy systems. Some industries must have sustainable production even during a power outage, which requires backup energy systems with adaptive power management to meet price and load variations with minimum impacts on grid. In addition, most of the distribution centers are associated with trucks. Transitioning into electric trucks is widely planned to reduce the negative impacts of transportations into climate change. There are a number of possible ways to implement fast-charging stations to support industrial facilities and the associated trucks.

This section explains the planning stage for implementing fast-charging station as integrated with a power grid substation. The location of the fast-charging station will be selected near industrial facilities and distribution centers where e-trucks are used for loading and unloading goods for distribution and supply chain. This scenario is selected to maximize the benefits of the fast-charging station to meet charging loads by e-trucks while supporting loads of distribution centers and other industrial facilities. The planning of fast-charging station should support normal operation and emergencies, including severe weather conditions. The number of trucks in Ontario is 131,952 and in Canada is 481,182. There are increasing moves to electrify these trucks as electric trucks or e-trucks.

Large distribution industries, such as department stores, wholesale stores, and other distribution companies, have an increase in electric loads due to factory automation systems. They are planning to have an enhanced energy system to support local loads with dynamic management with load shifting to enhance power flow and reduce cost of energy, in addition to stable operation and reliable services. Many industries are planning the transition to electric trucks with fast-charging capabilities to achieve minimum waiting times, longer driving ranges, and higher return on investment (ROI). There are a number of potential fast-charging station designs based on different energy storage technologies such as flywheel, ultracapacitor, and battery storage systems. This section will discuss the planning of fast-charging stations using hybrid energy storage including flywheel and battery systems.

The design requirements of the proposed fast-charging station include the following:

– Meet the capacity market.
– Achieve the target demand response.
– Meet the target voltage and frequency regulation requirements.

Charging stations are categorized into four main groups as follows:

1. Private charging stations
2. Public stations
3. Fast or quickly charging stations
4. Swapping or changing of battery process

Flywheel energy storage offers high power and energy density, which are suitable for fast-charging stations. Flywheel has high reliability with long lifecycle compared with the battery. Flywheel has no hazardous substances and can be recycled. Also, flywheel is less sensitive to temperature compared with batteries. Flywheel energy storage systems utilize active kinetic rotation that is stored with minimum friction losses. The input electrical energy is used to increase the acceleration and rotation of the mass. The stored energy will depend on the inertia and the rotating shaft. The optimum energy can be achieved by spinning the flywheel with the maximum possible speed.

The planned fast-charging station includes hybrid flywheel and battery storage and is called FFCS: flywheel-based fast-charging station. FFCS comprises advanced battery technologies with state-of-the-art flywheel technology to maximize the performance of the charging station in terms of demand response, voltage, and frequency regulation to meet customer requirements. Flywheel technologies can provide a fast and dynamic response to enhance the performance of the integrated battery storage within the stations. In addition, the flywheel can provide improved fast-charging capabilities to maximize the performance of battery storage by reducing the charging time. The integration of battery systems will provide a balance between the loads and supply while maximizing the overall station performance. This requires dynamic storage systems to charge the battery quickly and stabilize the dynamic behavior with charging/discharging cycles. The hybrid energy storage system will be integrated with the monitoring and control system as a unified operational solution. The integrated energy storage system will include advanced resiliency features to ensure sustainable and safe operation during normal and emergency conditions. FFCS will have comprehensive reliability and safety features in view of national standards and regulations from CSA (Canadian Standards Associations)/ IESO (Independent Electricity System Operator), with complete validations to ensure acceptable risk levels of the operation in all possible scenarios.

FFCS will support transportation electrification infrastructures for e-trucks, EVs, and e-buses. In addition, FFCS will support industrial facilities and power networks to meet local demands. The design of FFCS will be optimized in view of sizing of components and control strategies and operation schemes. The optimized performance will reduce operational costs and overall reductions in health-safety-environmental risks. The energy storage management system will provide resilient features to ensure minimum operation interruptions and higher availability. This is achieved using self-healing and fault-tolerant control mechanisms that can effectively decide the charging/discharging of the battery units and their integration to the flywheel and loads. Flywheel energy storage system can be considered the unique solution for the grids with peak demand and consumption due to its technical merits for energy profile with the advantage of the utilization of the mechanical energy and its conversion to electrical power.

The planning of these deployment projects will include the following main steps:

- Design FFCS with intelligent control system and power electronics; design integrated energy storage models using battery systems and flywheel technologies based on design requirements, load profiles, and other system specifications.
- Design energy management to integrate FFCS with industrial facilities and distribution centers with demand-side management schemes.
- Design interface system with e-trucks, charging/discharging, V2G (vehicle to grid), V2V, and G2V (grid to vehicle).
- Design energy management of FFCS to include mobility profiles, demand-side management, and power analysis for e-trucks.
- Define configuration parameters and optimize for the most suitable operational scenarios.
- Develop comprehensive economic and technical performance indicators for design and operation of the proposed integrated system.
- Evaluate risks and design fault-tolerant and self-healing approaches to ensure resilient energy systems.
- Evaluate and optimize the overall performance, including costs, quality, resiliency, and risks with environmental consideration for the potential design alternatives and control scenarios.
- Design and optimize control architecture with management system for integrated energy storage platform.
- Design and demonstrate the integrated energy storage system for a pilot site at one of the Loblaw distribution centers, which will be applied on a number of deployment projects, including utility grid, other residential, industrial facilities, and transportation electrification that are supplied.
- Assess and review the compliance with national and international standards and regulations, as well as regional energy policies
- Validate and verify the integrated energy storage system, and test and evaluate operational data and tuning of the pilot energy storage platform to quantify performance metrics in preparation for wider deployment and installation of FFCS.

The integration between battery technology and flywheel systems can offer high-performance and resilient energy storage platform that can deal with dynamic load profiles, quick and adaptive demand response, and high-performance charging/discharging mechanisms. The schematic of FFCS implementation is shown in Fig. 13.9.

FFCS structure includes flywheel energy storage platform (FESP), an integrated system to manage flywheel energy storage. FFCS includes resilient control

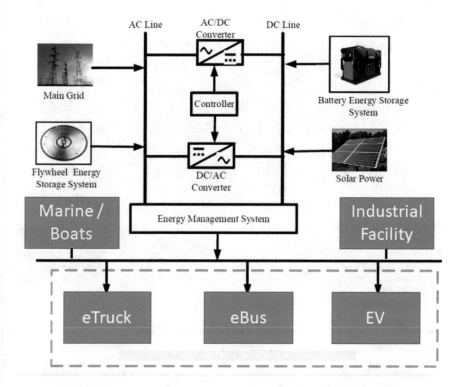

Fig. 13.9 Schematic of FFCS implementation

architecture (RCA), fast-charging controller (FCC), smart energy management and automation (SEMA), battery units (BU), and control and communication signals (C1, C2, etc.), as shown in Fig. 13.10.

The planning of FFCS implementations will focus on the evaluation and optimization of design and operation scenarios of FFCS and applications on industrial facilities, distribution centers, marine transportation electrification, and power substations, as shown in Fig. 13.11.

The integrated resilient control architecture is essential to support the different deployments for transportation electrification, substations, and industrial facilities. The proposed resilient control architecture is shown in Fig. 13.12, which includes control functions for power protection, power management, self-healing, resiliency, and supervisory control.

As part of the planning of FFCS deployment, existing trucks and trailers require modifications with retrofit solutions with integrated charging capabilities. Operating scenarios of e-trucks should include charging/discharging, V2G, V2V, and G2V options. The integration with the grid should be appropriately planned to ensure minimum impacts with different charging loads from transportation electrification. Different components' sizes and technology types within the FFCS will be planned as per the deployment strategy and application requirements. FFCS components

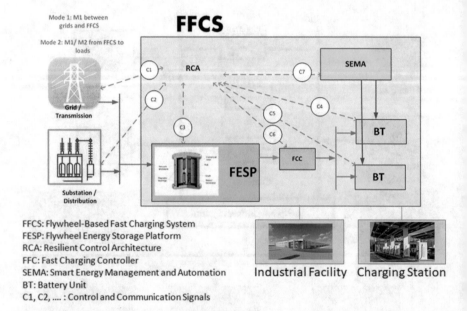

FFCS: Flywheel-Based Fast Charging System
FESP: Flywheel Energy Storage Platform
RCA: Resilient Control Architecture
FFC: Fast Charging Controller
SEMA: Smart Energy Management and Automation
BT: Battery Unit
C1, C2, ... : Control and Communication Signals

Fig. 13.10 FFCS hybrid structure for multipurpose implementations

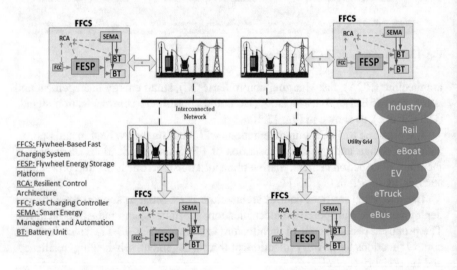

FFCS: Flywheel-Based Fast
Charging System
FESP: Flywheel Energy Storage
Platform
RCA: Resilient Control
Architecture
FFC: Fast Charging Controller
SEMA: Smart Energy
Management and Automation
BT: Battery Unit

Fig. 13.11 Deployment strategies of FFCS

will be optimized based on technical and economic KPIs in view of design and operation scenarios. The location of each FFCS will be analyzed and optimized as per different performance indicators. Study the replacement of battery storage in each e-truck to meet fast-charging requirements and target performance in different operating conditions. Energy management of the whole FFCS and for each truck

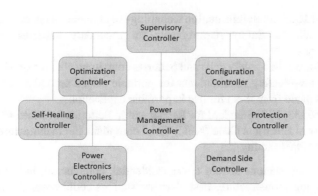

Fig. 13.12   Resilient control architecture (RCA) for FFCS implementations

will be planned in view of mobility and transportation routes. FFCS monitoring, control, and optimization will be evaluated with simulation models, including hardware-in-the-loop and real-time simulation.

Lifecycle-based performance assessment will be evaluated for each design and operation scenario of the FFCS while considering different charging profiles. The analysis will include load profiles of the industrial facility and distribution centers so that energy management can be planned accordingly. Resiliency and backup power supply will be planned as per target reliability and risk reduction expectations to meet safety levels. Stable power supply in normal and emergency situations will be evaluated to meet the target performance measures.

The planning will include the analysis of existing truck specifications to understand charging mechanisms and load profiles based on planned trips. The onboard energy storage system design in trucks will be evaluated and optimized based on different energy storage technologies.

The analysis of both off-grid and grid-connected scenarios will be evaluated as part of the planning of FFCS deployment with the considerations of integrated hybrid energy systems from renewable energy technologies. The station site will be evaluated in terms of location, layout, and possible paths to traffic and roads.

Resiliency analysis will be conducted to assess outage and failure scenarios and possible resiliency and protection layers. Self-healing and fault-tolerant design configurations and operation strategies will be planned to ensure sustained and stable energy supply and power flow within the FFCS, industrial facility, and trucks. FFCS modules will follow and comply with national and international standards, such as IEC (International Electrotechnical Commission), IEEE (Institute of Electrical and Electronics Engineers), and CSA, as follows:

– IEC 61508, including E/E/PE system: IEC 61508 (Parts 1–7) functional of electrical/electronic/programmable-related systems
– IEC 61427-2: For energy storage, cells, and batteries with general requirements and test methods
– IEC 61850: Communication networks and systems in substations

- IEEE Std 1625-2008: Information technology equipment—general requirements
- NF EN 50272: Safety requirements for secondary batteries and battery installations
- NF EN 61982: Secondary cells and batteries containing alkaline or other nonacid electrolytes— safety requirements for portable sealed secondary cells and for batteries made from them for use in portable applications
- NF EN 62133: EEE Standard for Rechargeable Batteries for Multi-Cell Devices
- US National Electrical Code (NEC) 690.5 compliant and manufactured in a certified ISO 9001 facility

Fast-charging stations should cover residential, commercial, industrial, and/or public charging infrastructure. Fast-charging stations could cover the majority of total electric demand. The increase of FFCS installations will lead to better utilization of energy supply and charging resources, with improved control of the peak demands. DC fast chargers will replace Level 1 and Level 2 charging stations. Fast charging can be achieved using Level 2 (basic) or Level 3 (ultimate), as below:

AC Level 1: 117V 16A maximum (normal charging)
AC Level-2: 240V 32A or 70A (basic fast charging)

Direct DC Level 3 by FFCS (ultimate):

- Input three-phase 200V
- Output: Maximum DC 45–50kW (which can reach 200kW), maximum DC voltage 700V, and maximum DC current 750A

Fast charging using FFCS will support congested distribution networks in large cities with peak demand reduction. Fast charging with FFCS will be able to transfer peak load into baseload.

The charging time using FFCS will be shorter than normal charging in the order of 3–6 times. For X kWh of charging in time t, the amount of normal charging $= X$ kWh, and the amount of FFCS charging $= 3X$ kWh. FFCS will offer reduced charging time which will benefit the impact on the grid as well as benefits of time savings for each charging load.

FFCS will cover most of the limitations in current charging technologies, which use battery, in terms of energy supply capacity, lifecycle cost, reliability, needs for frequent battery replacement, efficiency, maintenance, lifetime, operating temperature ranges, and dynamic responses. FFCS can enhance the congested distribution networks' performance while reducing peak demands.

In addition, FFCS will offer minimum frictional losses by utilizing advanced material of carbon-fiber-reinforced plastic, has noncontact bearings, has ultimate strength and speed up to 45,000 revolutions per minute, has high lifetime, and can range from the level of kW to the level of multi-MWs. FFCS offers a hysteresis control strategy that will increase its lifetime, interactively adjust the active power consumption, and regulate its power based on predicting the variation of production and load consumption. The integration between batteries and flywheels can solve a number of limitations that face each technology separately. The design and control of the integrated energy storage platform that comprises battery and flywheel technologies will support the deployment and implementation of FFCS in transportation electrification, commercial, and industrial facilities. The planning of FFCS deployment projects will include the following main tasks:

Task 13.1 Demands (distribution center and e-trucks), power flow, and operation
Task 13.2 Battery subsystem design and operation
Task 13.3 Flywheel subsystem design and operation
Task 13.4 Control system
Task 13.5 Performance evaluation and optimization
Task 13.6 Risk management, validation, and verification
Task 13.7 Data communication and security
Task 13.8 System integration

1. *FFCS Business Model*

The planning of FFCS deployments will include analysis of FFCS business model and identifying main requirements of the identified clients, such as end users, power supply, government regulations, and technology providers. This requires analysis of all requirements and process models and associated KPIs. This is essential to formulate reliable and realistic energy supply models within substations.

2. *FFCS Modeling and Simulation*

Integrated modeling and simulation with real-time simulation and hardware-in-the-loop systems will be utilized to evaluate different design and operation scenarios of the target FFCS based on the selected site and user requirements. It includes business process modeling to identify user requirements and map to grid physical and operational models, including reliability, integrity, control, and protection models.

3. *Multidimensional Optimization with KPIs*

The target FFCS control and protection will be optimized based on KPIs that are identified for each substation block and aggregated for the overall energy grid. Also, the demand prediction will allow cost optimization by balancing energy generation and supply to the grid (e.g., via distribution lines from utilities).

4. *Control and Protection System Design*

Distributed control strategies will be analyzed, and control system design based on intelligent adaptive control techniques will be developed using computational intelligence approaches, such as neuro-fuzzy and GA. The protection systems will be designed based on safety design concepts and independent protection layers (IPL) where inherent safety concepts will be analyzed to reduce cost and enhance the reliability of target grid protection systems. Distributed control structure based on a networked hierarchical control chart (HCC) will be developed and automated based on grid designs. The embedded control logic will be synthesized based on control rules using fuzzy logic controllers. Protection system design based on risk-based safety design will be employed to specify safety functions and map to safety systems.

5. *Real-Time Data Acquisition and Monitoring*

This task will include designing and implementing data acquisition systems, smart metering, monitoring, and communication technologies.

13.10.1 Deployment Impacts

The deployment of FFCS will reduce operational costs and GHG emissions for transportation and energy infrastructures. There are economic benefits to utilities, transportation sectors, industrial facilities, and end users. The deployment projects will include investment planning to expand into new business opportunities and commissioned projects. The deployment of FFCS will add value to existing power substations by merging industrial clusters of clean technology where the city will have competitive advantages with international opportunities.

The penetration of FFCS deployments will reduce energy consumption by supplying some demands from the energy storages systems with reduced charging time. For example, 5 MW FFCS could be used to support power substation and to operate as a backup power source. The cost of electricity set by the Ontario Energy Board is 8.3 ¢/kWh off-peak, 12.8 ¢/kWh mid-peak, and 17.5 ¢/kWh at on-peak. For simplicity, it is considered at 10 ¢/kWh. Hence, the 5 MWh would generate revenue of around $4.38M/year; considering timeframe as five years, the power savings is around $20M.

FFCS will have resilient intelligent controllers to achieve high performance and ensure minimum energy costs. FFCS will have optimum system configurations to increase efficiency, power availability, flexibility, dispatchability, and profits. FFCS will include real-time safety verification and risk monitoring of all components of the FFCS. The deployments of FFCS will lead to reduced greenhouse gas (GHG) emissions and will offer significant investment and profit.

The deployment of FFCS will lead to the following impacts in the transportation sector:

- Increase the availability and utilization of zero-emission vehicles on the road by deploying electric transportation.
- Increase the availability and use of fast chargers for vehicles, trucks, buses, rail, and marine transportation.
- Support accelerated construction of electric fast trains.

In the residential sector, the deployment of FFCS will lead to the following impacts:

- Reduce emissions in buildings and communities with more affordable charging infrastructures for homeowners, residents, and business owners to install or retrofit clean energy systems.
- Businesses and individuals will receive more information about better energy use and will be eligible for programs to encourage increased energy efficiency.

- Improve energy efficiency in multi-tenant residential buildings.
- Set lower carbon standards for communities with the expansion of fast-charging infrastructures.
- Improve energy efficiency in schools and hospitals.
- Promote low carbon energy supply and products.
- Help individuals and businesses manage their energy use and save money.

In the industrial sector, the following are potential impacts:

- Lead to clean manufacturing jobs.
- Balance industrial emission reductions.
- Continue economic competitiveness to help industries adopt low-carbon technologies.

And in terms of over FFCS performance, the following are key features:

- Provide stable, clean, cheap energy supply with battery and renewable energy integration for the operation of industrial facilities.
- Provide excess energy back to the grid.
- Incorporate resilient battery management to support energy supply to mechanical, electrical, and chemical systems, as well as the operation of industrial facilities.
- Provide resilient intelligent controller for high-performance control with optimization to ensure minimum energy costs, conservation, and high profit with improved ROI.
- Offer innovative, sustainable technology for balancing electricity supply and demand.
- Offer novel features with integrated energy storage platform and scalable flywheel-battery technologies.

13.11   Summary

This chapter discussed the planning of fast-charging stations as integrated with different infrastructures in view of design and operation parameters. The proposed planning approach will enable cities and municipalities to effectively plan for the deployment of fast-charging stations to achieve high performance at individual points based on integrated loads, system parameters, and user requirements. Discussions are presented on planning options to integrate fast-charging station for transportation electrification, power utility substation, and industrial facilities. Design and operation parameters are presented to show the best ways to plan the deployment by combing different options to maximize the overall performance and ROI of fast-charging stations.

References

1. S. Pelletier, O. Jabali, J.E. Mendoza, G. Laporte, The electric bus fleet transition problem. Transp. Res. Part C Emerg. Technol. **109**(October), 174–193 (2019)
2. L. Mathieu, Electric Buses arrive on time – Marketplace, economic, technology, environmental and policy perspectives for fully electric buses in the EU (2018)
3. M. Rogge, E. Van Der Hurk, A. Larsen, D.U. Sauer, Electric bus fleet size and mix problem with optimization of charging infrastructure. Appl. Energy **211**, 282–295 (2018)
4. Z. Wu, F. Guo, J. Polak, G. Strbac, Evaluating grid-interactive electric bus operation and demand response with load management tariff. Appl. Energy **255**(August), 113798 (2019)
5. M. Bhaskar Naik, P. Kumar, S. Majhi, Smart public transportation network expansion and its interaction with the grid. Int. J. Electr. Power Energy Syst. **105**(December), 365–380 (2019)
6. Y. Lin, K. Zhang, Z.-J.M. Shen, B. Ye, L. Miao, Multistage large-scale charging station planning for electric buses considering transportation network and power grid. Transp. Res. Part C Emerg. Technol. **107**(August), 423–443 (2019)
7. Y. He, Z. Song, Z. Liu, Fast-charging station deployment for battery electric bus systems considering electricity demand charges. Sustain. Cities Soc. **48**(October), 2019 (2018)
8. The European electric bus market is charging ahead, but how will it develop? | McKinsey [Online]. Available: https://www.mckinsey.com/industries/oil-and-gas/our-insights/the-european-electric-bus-market-is-charging-ahead-but-how-will-it-develop. Accessed 05 Mar 2020
9. A.B.B. Canada, Electrification of public transport. Toronto (2018)
10. Transport for London, Key bus routes in central London, 2020 [Online]. Available http://content.tfl.gov.uk/bus-route-maps/key-bus-routes-in-central-london.pdf. Accessed 06 Mar 2020
11. D. Perrotta et al., Route planning for electric buses: a case study in Oporto. Procedia – Soc. Behav. Sci. **111**, 1004–1014. Feb. 2014IRENA, "Innovation outlook: Smart charging for electric vehicles," 2019
12. M. Rogge, S. Wollny, and D. Sauer, "Fast charging battery buses for the electrification of urban public transport—a feasibility study focusing on charging infrastructure and energy storage requirements," Energies, vol. 8, no. 5, pp. 4587–4606, May 2015.
13. M. Andersson, Energy storage solutions for electric bus fast charging stations cost optimization of grid connection and grid reinforcements, Uppsala Universitet (2017)
14. Energy prices and costs in Europe | Energy [Online]. Available: https://ec.europa.eu/energy/en/data-analysis/energy-prices-and-costs. Accessed 06 Mar 2020
15. Q. Kang, J. Wang, M. Zhou, and A.C. Ammari, "Centralized charging strategy and scheduling algorithm for electric vehicles under a battery swapping scenario," IEEE Trans. Intell. Transp. Syst., vol. 17, no. 3, pp. 659–669, Mar. 2016.
16. H.A. Gabbar, A.M. Othman, Flywheel-based fast charging station – FFCS for electric vehicles and public transportation. IOP Conf. Ser. Earth Environ. Sci. **83**, 012009 (2017)

Chapter 14
Techno-economic Analysis of Fast-Charging Infrastructure

Hossam A. Gabbar

Nomenclature

$EV_{ct}(i)$	Average EV charging time at fast-charging station (i) (h)
$EV_{cm}(i,j)$	Charging mode of EV (i) at trip (j) ("R," "F," "B")
$EV_{dt}(i,j)$	Daily trip of EV (i) (km)
$EV_{nt}(i,j)$	Night trip of EV (i) (km)
$EV_{dc}(i,j)$	Daily charged EV at station (i) (number)
EV_{bs}	Average EV battery size (kWh)
EV_{e-avg}	Average energy consumption for EV (kWh/km)
$EB_{ct}(i)$	Average e-bus charging time at fast-charging station (i) (h)
$EB_{cm}(i,j)$	Charging mode of EB (i) at trip (j) ("R," "F," "W," "B")
$EB_{dt}(i)$	Daily trip of EB (i) (km)
$EB_{nt}(i)$	Night trip of EB (i) (km)
$EB_{dc}(i,j)$	Daily charged EB at station (i) (number)
EB_{bs}	Average EB battery size (kWh)
$EM_{ct}(i)$	Average e-boat charging time at fast-charging station (i) (h)
$EM_{cm}(i,j)$	Charging mode of EM (i) at trip (j) ("R," "F," "W," "B")
$EM_{dt}(i)$	Daily trip of EM (i) (km)
$EM_{nt}(i)$	Night trip of EM (i) (km)
$EM_{dc}(i,j)$	Daily charged EM at station (i) (number)
EM_{bs}	Average EM battery size (kWh)
$ET_{ct}(i)$	Average e-truck charging time at fast-charging station (i) (h)
$ET_{cm}(i,j)$	Charging mode of ET (i) at trip (j) ("R," "F," "W," "B")

This chapter is contributed by Hossam A. Gabbar.

H. A. Gabbar (✉)
Energy Systems and Nuclear Science, University of Ontario Institute of Technology, Oshawa, ON, Canada
e-mail: Hossam.gabbar@uoit.ca

© Springer Nature Switzerland AG 2022
H. A. Gabbar, *Fast Charging and Resilient Transportation Infrastructures in Smart Cities*, https://doi.org/10.1007/978-3-031-09500-9_14

$ET_{dt}(i)$	Daily trip of ET (i) (km)
$ET_{nt}(i)$	Night trip of ET (i) (km)
$ET_{dc}(i,j)$	Daily charged ET at station (i) (number)
ET_{bs}	Average ET battery size (kWh)
FCS_n	Number of fast-charging stations (number)
$SS_p(i)$	Position of substation (i) (location x,y)
$IF_p(i)$	Position of industrial facility (i) (location x,y)
$FCS_p(i)$	Position of fast-charging station (i) (location x,y)
$FCS_s(i)$	Fast-charging station (i) serving land, marine, substation, or industrial facility ("land," T/F; "marine," T/F; "substation," T/F; "industry," T/F)
$FCS_{dl}(i)$	Total daily load at fast-charging station (i) (kWh)
$FCS_{ul}(i)$	Maximum load at fast-charging station (i) (kWh)
$FCS_{ll}(i)$	Minimum load at fast-charging station (i) (kWh)
$FCS_{dm}(i)$	Distance from fast-charging station (i) where battery is swapped to main charging station (j) for battery charging (km)
$SS_{dl}(i,j)$	Daily load at substation (j) supported by fast-charging station (i) (kWh)
$IF_{dl}(i,k)$	Daily load at industrial facility (k) supported by fast-charging station (i) (kWh)

14.1   Introduction

Based on the 2030 Agenda for Sustainable Development, there are seventeen Sustainable Development Goals (SDGs) planned by both developed and developing countries. It is essential to provide affordable and clean energy [1]. This includes the transportation sector, a significant source of greenhouse gas emissions [2]. Transportation electrification can be achieved by establishing robust charging infrastructures with capabilities to meet all public, utility, and municipality requirements. Technology development of fast-charging stations and the improvements in bus technologies enabled wider deployment.

Similarly, the technology development of electric trucks (ETs) is mobilizing the implementation of fast charging to support a longer range of delivery. The techno-economy analysis effectively evaluates design and operation options for transportation electrification [3]. A practical framework is applied on charging electric buses (EBs) with different charging models [4]. The charging models require a transformation of the bus fleet, as explained in [5]. Optimizing the charging infrastructure is essential for profitable deployment projects [6]. The optimization should be implemented based on different types of buses [7]. The cost model could be studied in view of demand on the grid side [8]. The charging infrastructure planning is optimized by considering transportation network and grid interactions [9]. The planning could consider charging ahead to better understand transit and mobility models [10]. Energy storage requirement is studied for fast-charging infrastructures [11]. Considering energy storage, cost optimization could be achieved for charging stations [12]. The planning is implemented on bus network with different route requirements [13, 14]. Charging scheduling is analyzed for different scenarios of transportation network [15]. Flywheel-based fast-charging station is proposed as one key technology for improved performance of charging technologies [16]. The location of fast-charging infrastructure is optimized with AI (artificial intelligence) algorithms [17].

This chapter will discuss the techno-economic analysis of fast-charging infrastructures based on possible strategies and technical specifications, as explained in the following sections.

14.2 Integrated Deployment Model of Fast-Charging Stations

The integrated deployment of fast-charging stations is presented in Fig. 14.1. The fast-charging infrastructure will support EVs (electric vehicles), electric buses, electric trucks, and electric boats while supporting substations and industrial facilities.

There are a number of scenarios that could be realized using the integrated deployment model. These scenarios are based on design and operation configurations variations as per Table 14.1.

The framework of estimating daily loads at a given FCS (fast-charging system) is shown in Fig. 14.2. It includes the daily load calculation due to local loads at the FCS station and loads to support swapped batteries from other nearby stations. When applicable, the daily load at each station will include charging EVs, EBs, ETs, and EMs. Also, if the fast-charging station supports the substation, it will include charging loads. Similarly, if the fast-charging station supports an industrial facility, such as a distribution center, the total loads will include charging loads from the industrial facility.

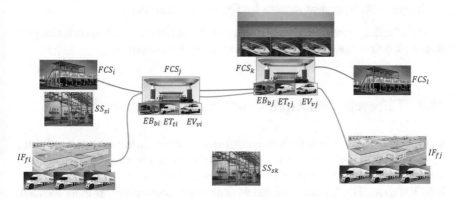

Fig. 14.1 Integrated deployment of fast-charging station

Table 14.1 Scenario model parameters of the integrated deployment of fast-charging station

Main item	Element	Option
Fast-charging station	S1 serving	S1.1 Land (EV, EB, ET), $FCS_s(i)$ = "L"
		S1.2 Beside shore (EV, EB, EM, ET), $FCS_s(i)$ = "M"
		S1.3 Beside substation (EV, EB, ET, SS), $FCS_s(i)$ = "S"
		S1.4 Beside industrial facility (EV, EB, ET, IF), $FCS_s(i)$ = "I"
EV	S2 charging mode of EV (i) for trip (j)	S2.1 Regular charging, $EV_{cm}(i,j)$ = "R"
		S2.2 Fast charging, $EV_{cm}(i,j)$ = "F"
		S2.3 Wireless charging, $EV_{cm}(i,j)$ = "W"
		S2.4 Battery swap, $EV_{cm}(i,j)$ = "B"
EB	S3 charging mode of EB (i) for trip (j)	S3.1 Regular charging, $EB_{cm}(i,j)$ = "R"
		S3.2 Fast charging, $EB_{cm}(i,j)$ = "F"
		S3.3 Wireless charging, $EB_{cm}(i,j)$ = "W"
		S3.4 Battery swap, $EB_{cm}(i,j)$ = "B"
EM	S4 charging mode of EM (i) for trip (j)	S4.1 Regular charging, $EM_{cm}(i,j)$ = "R"
		S4.2 Fast charging, $EM_{cm}(i,j)$ = "F"
		S4.3 Wireless charging, $EM_{cm}(i,j)$ = "W"
		S4.3 Battery swap, $EM_{cm}(i,j)$ = "B"
ET	S5 charging mode of ET (i) for trip (j)	S5.1 Regular charging, $ET_{cm}(i,j)$ = "R"
		S5.2 Fast charging, $ET_{cm}(i,j)$ = "F"
		S5.3 Wireless charging, $ET_{cm}(i,j)$ = "W"
		S5.4 Battery swap, $ET_{cm}(i,j)$ = "B"

Total daily load to charge swapped batteries = from current station (number of batteries from EVs * average EV battery size + number of batteries from EBs * average EB battery size + number of batteries from EMs * average EM battery size + number of batteries from ETs * average ET battery size) + $\sum$ batteries charged from other near stations for EV, EB, ET, EM) (kWh)

The total daily load will be used to estimate the optimum size of the fast-charging station as well as the cost analysis of the charging infrastructure.

14.3   Lifecycle Cost Analysis of Charging Station

The lifecycle cost analysis is an important tool to assess different scenarios for the deployment of fast-charging station at a given location. The lifecycle cost analysis includes the options to support substation, industrial facility, and support to EVs, EBs, EMs, and ETs. The analysis should demonstrate maximized profit for the identified scenario. The analysis will include all expenses that are compared with the expected profits from the station, as shown in Table 14.2.

Fig. 14.2 Framework to calculate total daily load at FCS

14.3.1   Cost Calculation

In order to demonstrate cost calculation, one FCS is installed beside the substation, where FCS design is specified to meet target load profile of the substation to maximize profit. The size of the FCS is optimized to meet the target load profile. Based on the load profile, the following details are estimated:

FCS: 1 MW, 3794.7 MWh
Flywheel: 1 MW
Battery: 1 MW

For example, to support demand activities at the target substation and to operate as a backup power resource for existing units, the costs of electricity set by the Ontario Energy Board are the following:

8.3¢/kWh off-peak
12.8¢/kWh mid-peak
17.5¢/kWh at on-peak

The above costs will be used to estimate ROI accordingly.

Total power saving = 5.27 * 10³ kW = 5.27 MW (per month)
Total energy saving = 3744.68 MWh (per month)

Table 14.2 Cost analysis of the integrated deployment of fast-charging station

Main item	Element	Details
FCS infrastructure cost	Capital cost	FCS size MWh
		Flywheel, size MW
		Battery, size MW
		Interface to substation
		Interface to industrial facility
		Number of FCS installations
	Annual operating cost	Facility cost
		Transportation cost of swapped batteries
		Energy supply cost:
		Off-peak
		Mid-peak
		On-peak
Bus	Bus technology	Bus type
		12-meter
		18-meter or articulated
		24-meter or double articulated
	Bus trips	Bus route
		Number of trips (per day)
		Number of stops (per trip)
		Total number of stops (per day)
		Trip length (km)
		Bus size (m)
		Average consumption (kWh/km)
Truck	Truck technology	Truck type (engine size)
		Model-T1
		Model-T2
		Model-T3
	Truck trips	Truck route
		Number of trips (per day)
		Average load weight per trip
		Trip length (km)
		Truck size (m)
		Average consumption (kWh/km)
Boat	Boat technology	Boat type (engine size)
		Model-B1
		Model-B2
		Model-B3
	Boat trips	Boat route
		Number of trips (per day)
		Average load weight per trip
		Trip length (km)
		Boat size (m)
		Average consumption (kWh/km)

Total saving = \$157.27 * 10^3 (per month)
Annual total saving = \$1320.48 * 10^3

The corresponding ROI (return on investment) for both options, (1) fast-charging station with battery system only and (2) fast-charging station with flywheel, are as below:

	Battery system	FCS
Total capital cost	\$1,200,000	\$3,800,000
Total operating cost/year	\$80,000	\$130,000
Annual energy saving	\$330,220	\$1,320,480
ROI in years	$4.8 \approx 5$	$3.2 \approx 4$

Comparing both options, (1) FCS versus the battery system is based on FCS that can offer four times faster charging time than a typical battery system.

For monthly charging demand of 3744.68 MWh, with 1.5 h charging time:

- The energy amount provided by the battery system = 936.57 MWh
- The energy amount provided by the FCS = 3744.68 MWh

Considering the time of FCS for the same amount of 3744.68 MWh of charging demand, the charging time of the battery system is 4–6 times longer than the time of FCS.

14.4 Techno-economic Analysis

In order to illustrate the techno-economic analysis of the fast-charging station deployment, selected scenarios are evaluated using simulation models to evaluate key performance indicators and select the best deployment strategy. The list of scenarios is defined in Table 14.3, where the main scenario and associated attributes are defined, along with a list of key performance indicators (KPIs) to evaluate each scenario.

The analysis is conducted based on the selected values as shown in Table 14.4 below.

Table 14.3 Scenarios for the selected case study

Scenario	Attributes	KPIs
S1. FCS1 supports charging EV, EB, ET, and EM	Daily charged EV, EB, EM, and ET	KPI1.1: annual profit KPI1.2: average wait time to charge KPI1.3 energy savings
S2. FCS1 supports charging EV, EB, ET, and EM and SS1	Daily charged EV, EB, EM, and ET Daily load at FCS by SS1	KPI2.1: annual profit KPI2.2: average wait time to charge KPI2.3 energy savings
S3. FCS1 supports charging EV, EB, ET, and EM and SS1 and IF1	Daily charged EV, EB, EM, and ET Daily load supplied by FCS for SS1 Daily load supplied by FCS for IF1	KPI2.1: annual profit KPI2.2: average wait time to charge KPI2.3 energy savings

Table 14.4 Scenario evaluation

Scenario	KPI	Evaluation
S1. FCS1 supports charging EV, EB, ET, and EM	KPI1.1: annual profit	Annual capital cost of FCS = (capital cost/lifetime) Annual operating cost of FCS = annual cost of electricity + annual operation cost + annual maintenance cost + annual labor cost Annual income = annual income from EV + annual income from EB + annual income from ET + annual income from EM Annual income from EV = daily EV charging count * 365 * EV yearly factor * EV one-time charging amount * rate per kWh (for each peak period) Annual income from EB = daily EV charging count * 365 * EV yearly factor * EB one-time charging amount * rate per kWh (for each peak period) Annual income from ET = daily ET charging count * 365 * ET yearly factor * ET one-time charging amount * rate per kWh (for each peak period) Annual income from EM = daily EM charging count * 365 * EM yearly factor * EM one-time charging amount * rate per kWh (for each peak period)
	KPI1.2: average wait time to charge	The model is defined for a single line multi-charger head. Λ: average arrival rate of EV, EB, ET, and EM μ: average charging rate of EV, EB, ET, and EM s: number of charging heads in the station $\mathcal{f}$: $\dfrac{\Lambda}{(s\mu)}$ average charger utilization P_0: $\left[\sum_{n=0}^{s-1}\dfrac{(\Lambda/\mu)^n}{n!} + \dfrac{\left(\frac{\Lambda}{\mu}\right)^s}{s!}\left(\dfrac{\mathcal{f}}{(1-\mathcal{f})}\right)\right]^{-1}$ = probability that no EVs are in the station L_Q: $\dfrac{\mathcal{f}_0\left(\frac{\Lambda}{\mu}\right)^s \mathcal{f}}{s!(1-\mathcal{f})^2}$ = average number of EV waiting in line W_Q: L_Q/Λ = average time spent waiting in line W: $W_Q + \dfrac{1}{(\mu)}$ = average time spent in the station, including charging L: ΛW = average number of EV in charging service P_n: $\begin{cases} \dfrac{\left(\frac{\Lambda}{(\mu)}\right)^n}{n!}P_0 \; for \, n \le s \\ \dfrac{\left(\frac{\Lambda}{(\mu)}\right)^n}{s!\left(s\right)^{n-s}}P_0 \; for \, n > s \end{cases}$ = probability that EVs are in station at a given time The same formula is applied on EB, ET, and EM, as per [18]
	KPI1.3 energy savings	Total annual amount of energy shifted from on-peak to mid-peak or off-peak + total amount of energy shifted from mid-peak to off-peak, due to the use of flywheel energy storage platform

(continued)

Table 14.4 (continued)

Scenario	KPI	Evaluation
S2. FCS1 supports charging EV, EB, ET, and EM and SS1	KPI2.1: annual profit	Same as scenario S1, while considering the total size of FCS to support SS1 loads
	KPI2.2: average wait time to charge	Same as scenario S1, while considering load served to SS1 during charging of EV, EB, ET, and EM
	KPI2.3 energy savings	Same as scenario S1, while considering energy savings at SS1 due to the use of FCS
S3. FCS1 supports charging EV, EB, ET, and EM and SS1 and IF1	KPI2.1: annual profit	Same as scenario S1, while considering the total size of FCS to support SS1 and IF1 loads
	KPI2.2: average wait time to charge	Same as scenario S1, while considering load served to SS1 and IF1 during charging of EV, EB, ET, and EM
	KPI2.3 energy savings	Same as scenario S1, while considering energy savings at SS1 and IF1 due to the use of FCS

14.5 Summary

There are several ways to implement fast-charging infrastructures in view of transportation loads and possible integration strategies. This chapter presented a number of scenarios to integrate fast-charging stations with power utility substations or industrial facilities, such as distribution centers. The proposed fast-charging infrastructures are analyzed for electric vehicles, buses, trucks, and boats (for marine transportation). Cost analysis is conducted based on modeling of different technologies and was applied on a case study where energy savings are achieved using fast-charging station with a substation.

References

1. O. Edenhofer et al., *Climate Change 2014 Mitigation of Climate Change Working Group III Contribution to the Fifth Assessment Report of the Intergovernmental Panel on Climate Change Edited by* (2014)
2. International Energy Agency, World energy balances overview (2019)
3. D. Nicolaides, A.K. Madhusudhanan, X. Na, J. Miles, D. Cebon, Technoeconomic analysis of charging and heating options for an electric bus service in London. IEEE Trans. Transp. Electrif. **5**(3), 769–781 (2019)
4. M. Lotfi, P. Pereira, N.G. Paterakis, H.A. Gabbar, J.P.S. Catalão, Optimal design of electric bus transport systems with minimal total ownership cost. IEEE Access, 1–16. Print ISSN: 2169-3536, Online ISSN: 2169-3536. https://doi.org/10.1109/ACCESS.2020.3004910
5. S. Pelletier, O. Jabali, J.E. Mendoza, G. Laporte, The electric bus fleet transition problem. Transp. Res. Part C Emerg. Technol. **109**(October), 174–193 (2019)

6. M. Rogge, E. Van Der Hurk, A. Larsen, D.U. Sauer, Electric bus fleet size and mix problem with optimization of charging infrastructure. Appl. Energy **211**, 282–295 (2018)
7. E. Yao, T. Liu, T. Lu, Y. Yang, Optimization of electric vehicle scheduling with multiple vehicle types in public transport. Sustain. Cities Soc. **52**(August 2019), 101862 (2020)
8. Y. He, Z. Song, Z. Liu, Fast-charging station deployment for battery electric bus systems considering electricity demand charges. Sustain. Cities Soc. **48**(October 2018) (2019)
9. Y. Lin, K. Zhang, Z.-J.M. Shen, B. Ye, L. Miao, Multistage large-scale charging station planning for electric buses considering transportation network and power grid. Transp. Res. Part C Emerg. Technol. **107**(August), 423–443 (2019)
10. The European electric bus market is charging ahead, but how will it develop? (McKinsey). [Online]. Available https://www.mckinsey.com/industries/oil-and-gas/our-insights/the-european-electric-bus-market-is-charging-ahead-but-how-will-it-develop. Accessed 05 Mar 2020
11. M. Rogge, S. Wollny, D. Sauer, Fast charging battery buses for the electrification of urban public transport – A feasibility study focusing on charging infrastructure and energy storage requirements. Energies **8**(5), 4587–4606 (May 2015)
12. M. Andersson, *Energy Storage Solutions for Electric Bus Fast Charging Stations Cost Optimization of Grid Connection and Grid Reinforcements* (Uppsala Universitet, 2017)
13. D. Perrotta et al., Route planning for electric buses: A case study in Oporto. Procedia–Soc. Behav. Sci. **111**, 1004–1014 (Feburary 2014)
14. Transport for London, Key bus routes in central London (2020). [Online]. Available http://content.tfl.gov.uk/bus-route-maps/key-bus-routes-in-central-london.pdf. Accessed 06 Mar 2020
15. Q. Kang, J. Wang, M. Zhou, A.C. Ammari, Centralized charging strategy and scheduling algorithm for electric vehicles under a battery swapping scenario. IEEE Trans. Intell. Transp. Syst. **17**(3), 659–669 (March 2016)
16. H.A. Gabbar, A.M. Othman, Flywheel-based fast charging station – FFCS for electric vehicles and public transportation. IOP Conf. Ser. Earth Environ. Sci. **83**, 012009 (August 2017)
17. A.M. Othman, H.A. Gabbar, F. Pino, M. Repetto, Optimal electrical fast charging stations by enhanced descent gradient and Voronoi diagram. Comput. Electr. Eng. **83**, 106574 (May 2020)
18. Waiting Line Models. https://www.csus.edu/indiv/b/blakeh/mgmt/documents/opm101supplc.pdf

Chapter 15
Advances in Charging Infrastructures

Hossam A. Gabbar

15.1   Introduction

Several factors influence public EV charging infrastructures [1]. The growth of charging stations and the development of EV technologies will support the penetration of EVs. Electric vehicles (EVs) and plug-in hybrid electric vehicles (PHEVs) are charged to serve mobility demands. However, with different electricity prices and tariffs, EVs can be considered as energy storage to shift peak from on-peak to off-peak prices. Also, EVs can be used to supply energy back to grid, homes, or facilities in emergencies or on demand. From one side, the introduction of EVs and PHEVs increases charging rates, negatively impacting the grid. However, they can be helpful to balance energy demand and support energy management. The concept of V2G or vehicle to grid will enable supply excess power from EV when not used back to the grid. The implementation of V2G is suitable for parking, malls, houses, farms, industrial facility, and buildings. Vehicle to vehicle (V2V) offered effective ways to charge EVs in different modes, which required establishing communication networks, such as ad hoc mobile wireless networks and protocols [2]. V2V charging allocation protocols using VANET (vehicular ad hoc network) are explained in [3]. V2V can be designed using single-phase EV on-board charger with charging assistance capabilities [4]. Mobile charging could be managed with online systems that will organize EVs' charging and discharging while communicating with both

This chapter is contributed by Hossam A. Gabbar.

H. A. Gabbar (✉)
Energy Systems and Nuclear Science, University of Ontario Institute of Technology,
Oshawa, ON, Canada
e-mail: Hossam.gabbar@uoit.ca

owners [5]. Similar concept is introduced as the Internet of vehicle as part of smart cities' planning [6]. Using proper analysis, all charging functions should achieve flexible and high grid performance [7]. Effective V2V energy sharing can be realized using matching mechanisms using techniques such as game theory while considering user requirements and satisfaction [8]. Cooperative approach can be utilized to enhance V2V charging functions [9]. The V2V charging should consider spatial resource constraints to achieve realistic charging performance [10]. The planning of V2V charging can be improved if considered traffic conditions where charging could occur during waiting regions [11]. Similarly, highway conditions can improve charging performance [12]. The planning of EV charging is analyzed within smart city outlook with good understanding of penetration and sales [13, 14]. The analysis of EV charging has been extended to discuss planning of fast-charging stations and infrastructures [15].

15.2   V2G Charging

15.2.1   V2G System Design

V2G system will be improved by integrating V2V capabilities with enhanced power transfer. Figure 15.1 shows the development framework of V2G where the system design shows the connections of PV, batteries, energy management system (EMS),

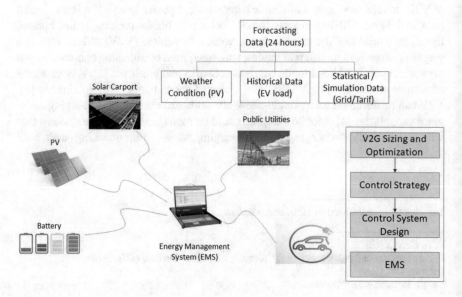

Fig. 15.1 Framework of V2G system design

and interface to the grid. For effective energy management, weather conditions are analyzed and mapped to 24-h forecasting, simulation, and historical data.

15.2.2   V2G Deployment

A hybrid charging station with renewable energy sources and battery storage systems can be installed in parking areas. Parking will be modified to maximize the area to install PV systems, such as parking roof. The design of the hybrid energy system and the integration with the charging system is analyzed in view of user requirements, charging demand profiles, and energy resources in the area. There are several performance measures to evaluate the design and operation of the V2G system, which are used to evaluate size, configuration, and control strategies while considering the optimum performance to the grid. Optimal power flow techniques are used to analyze the different components and their configurations and to optimize performance measures, including minimum energy costs, for example, operation and installation costs. The proposed V2G solution will maximize PV penetration to support the growing needs for EV charging. V2G system topology includes EV, energy storage, converter, and other battery storage systems. The target system configuration and operation should profit from the target V2G system, from selling energy back to grid and charging public EVs. The sizing of the hybrid V2G system can be optimized while maximizing the overall performance, including reducing the demand from the grid. To achieve effective energy management, AI (artificial intelligence) and intelligent algorithms are used to control power flow. User interface communicates input parameters and output measures with end users and indicates monitoring data and buying and selling process details. Another design feature is the utilization of the Open Charge Point Protocol (OCPP) that enables EV drivers to seamlessly access different charging networks across Canada and the United States through one network.

15.2.3   Benefits

V2G systems offer significant benefits to promote EV as a clean transportation system, in particular when countries have shortage of fossil fuel. V2G can offer reduced charging time of EVs via implementing fast-charging capabilities in the power circuit. The effective V2G solution will minimize power supplied from the grid, reducing system running costs. The design of V2G system is improved by optimizing the size of PV, battery, and power electronics, which will lead to reduced installation costs.

15.3   Control Strategy

The control optimization will consider SoC (system of charge) of the battery systems, which might occur due to different driving cycles, battery capacity, customer behavior, or weather. The analysis of SoC of different battery systems will be used to coordinate the charging process based on SoC level so that batteries with minimum SoC will be charged while utilizing batteries with higher SoC with charging. For example, if there are three EVs in the charging station with 60%, 20%, and 80% SoCs, customers can sell their energy capacity considering determined rates with other customers. In this case, we can use the largest SoCs 60% and 80% to charge the battery with the lowest SoC 20%. In this case, we can reduce the amount of power capacity of the hybrid charging system using customer EV batteries. The proposed V2G topology will offer to charge EVs from the neighborhood, which will reduce high demand on the grid.

The proposed energy management system will offer optimum performance to achieve minimum operating cost, minimum charging time, increased profits, reduced energy cost, enhanced grid performance, and improved system reliability, including battery lifetime.

The implementation of V2G solutions will offer a number of key functions to end users, including charging EV, supplying power back to the grid, managing buy and sell process, system protection, and reducing grid impacts.

15.4   V2G Installation

The installation process will start with site visits to collect data about the target parking area and study user requirements. The understanding of load profile will enable proper station design and sizing. The integrated design and interfaces are analyzed and evaluated while configuring operation parameters. Installation and testing of the integrated system include electrical, structural, PV, battery, charging unit, bidirectional smart meters, and power electronics. System configuration will ensure that design requirements are met and target performance is achieved. Monitoring functions will be installed within the user interface to collect real-time data and analyze performance. Maintenance and inspection tasks are defined and managed to ensure system integrity and safe operation while achieving long system life.

The most important performance measure for EV charging is the charging time. Various standardized charging levels are developed for charging at different power levels. The charging stations are categorized into three classes and are grouped as AC and DC charging stations. The summary of the different power and current levels on these standards for EV charging stations is analyzed in Table 15.1.

There are several key benefits of adopting V2G charging stations. There are a number of benefits to the grid, as follows:

– Support peak power.
– Turn peak load into baseload.

Table 15.1 EV charging standards

Levels	Maximum power rating (kW)	Maximum current rating (A)
IEC Standard		
AC charging		
Level 1	4–7.5	16
Level 2	8–15	32
Level 3	60–120	250
DC charging	100–200	400
SAE Standard		
AC charging		
Level 1	2	16
Level 2	20	80
Level 3	Above 20	Above 80
DC charging		
Level 1	40	80
Level 2	90	200
Level 3	240 and more	400
CHAdeMO		
DC fast charging	62.5	125

- Allow grid stability.
- Support grid expansion.
- Maintain promising supply levels.

Figure 15.2 shows the proposed system architecture of V2G system to support deployment in different sites and applications.

In order to ensure proper energy management, a detailed control strategy is defined and mapped to the V2G system design, as shown in Fig. 15.3.

15.5 Case Study V2G System Design

In order to understand V2G implementation, this section describes the selected components and detailed specifications.

Solar PV Panels
30 panels: FS270 W mono solar panel
Vmp: 30 V Voc: 36 V Imp 9 A
Size: 1640*992*40 mm
25-year lifetime
Glass: 3.2 mm tempered
Terminal block: IP65 with MC4 connector
Efficiency performance: 10 years 90%, 25 years 80%, and 20 years 85%

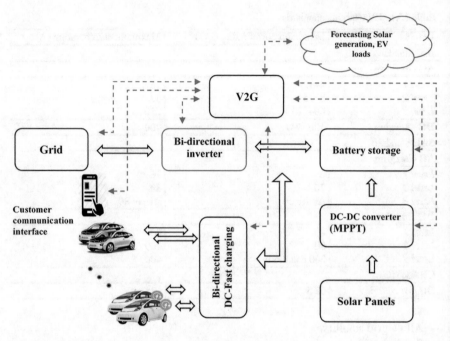

Fig. 15.2 Proposed system architecture of V2G station

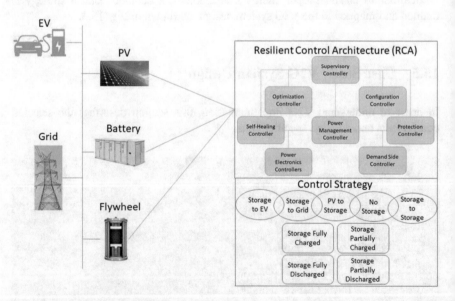

Fig. 15.3 V2G system design and control strategy

PV Array Combiner
Model: H6T-96v multiple PV string inputs
Simplify wiring between PV array and controller, protections to controller, and features
Wide range of DC input voltage
Reliable thunderstorm and surge protection

Hybrid Inverter
Inverter power: 8kw/96 V
MPPT controller model: 96v/80 A
Charging efficiency: 90–95%
LCD display inverter and controller info
Built-in **AC charger** (automatically)
O/P: 110 V, 220 VAC and 5 V/12 VDC
Product size: 590*470*730 mm

Battery Pack
Capacity: 12 V/150 AH
Full-sealed solar power gel battery
Service life: 6–8 years
Size: 483*170*219 mm

PV Panel Racks
Slope rooftop or flat rooftop
Wind load: 55 m/s, snow load: 1.5 kn/m^2
Structure: Anodized aluminum + stainless steel
Angle adjustable

Cables, Wiring, and Connections
1. 14 pcs 16 mm 2*35 CM battery cable
     4 pcs 16 mm 2*1 M battery cable
2. 4 mm 2 PV cable 150 M
3. Terminals and MC4.

15.6   Flywheel-Based Fast Charging

There are several innovations in the V2G design. For example, flywheel energy storage could provide higher power density than conventional battery systems. Higher power density will directly impact the overall charging performance. In addition, flywheel systems offer longer lifecycle compared with battery. Moreover, flywheel has higher reliability which will reduce operating costs. From the environmental impact side, flywheel has minimum negative impacts on the environment with non-hazardous materials and is easier to recycle than battery systems. Flywheel is less sensitive to temperature, while batteries are negatively impacted with high and low temperatures where their performance is reduced. The temperature sets some important thermal limits in the windings to avoid melting, in the magnets to avoid demagnetization, and in the composite materials to avoid burning it.

15.7   Case Study V2G with Commercial Building

A sample system integration within commercial buildings is shown in Fig. 15.4. Consider a commercial building with a parking area for 40 cars. We consider fast-charging station with a capacity of 20% of total vehicles, which is eight EVs. The estimated peak power limit for fast charging of eight EV batteries is approximately 100 kW × 8 = 800 kW. If this power is directly pulled from the grid, it will negatively affect it. Accordingly, the system design will include new equipment and infrastructure such as a transformer, cables, protection equipment, etc.

The proposed hybrid charging system will support incoming EVs as per parking space. In this case, one car parking space is 2.75 meters × 6 meters. To support 40 cars in the parking, the parking area is approximately 700 m². Also, we can use the roof of the building to implement solar panels. The building roof area is approximately 2000 m², with total PV panel area around 2500 m². PV power generation

Fig. 15.4 V2G deployment in commercial building

depends on the site location (weather, daylight time, and other location-related parameters). Considering the general amount of daylight time as 10 h, the PV generation is approximately 200 W for each m^2. Accordingly, total PV generation is 500 kW peak power generation. A battery storage system is 500 kW for hybrid system design and provides smooth output power. The proposed energy management system will manage charging power while supporting the hybrid system, minimizing the peak effect on the grid.

The initial investment cost of the proposed V2G system is expected to be high. Therefore, minimizing the initial investment costs of such systems with innovative solutions will provide an essential financial advantage. In addition, the offered V2G system will support municipality or state incentives. The economic benefits of the system will provide profitable installations.

15.8   Wireless Charging

There are two types to charge EV: (1) plug-in charging and (2) wireless charging. Plug-in charging has some challenges, such as the difficulty of charging cables, which present a potential trip hazard and degradation of use over time. Also, the manual operation of physically inserting and removing the charger is another challenge [16]. Wireless charging using wireless power transfer (WPT) can address these problems. WPT is based on magnetic coupling, which leads to inductive wireless power transfer (IPT). Several technology providers offer WPT technologies, such as plugless power [17]. Auto industries promote WPT, for example, Nissan LEAF, Chevy Volt, Cadillac ELR, BMW i3, Mercedes S550e, Tesla Model S, and other auto manufacturers. The typical power levels can reach 3.7 kW (WPT 1), 7.7 kW (WPT 2), and 11 kW (WPT 3) [18]. Wireless charging could be implemented directly from the station to EV or V2V, as shown in Fig. 15.5. V2V wireless charging can be done during driving based on V2V communications and agreement on V2V charging terms, spec, conditions based on SoC, and mobility and trip plan.

Static wireless charging is achieved when EV is charged while stopping. Dynamic wireless charging is achieved when EV is moving. Wireless EV charging can be improved to reach fast charging [19]. The fast charging capabilities will support a number of applications with specific mobility demands. However, it will require an upgraded circuit design to accommodate fast charging.

There are different coupling mechanisms. Inductive coupling wireless power transfer (IPT) and capacitive coupling wireless power transfer (CPT) are possible approaches with DC link to implement wireless EV charging. Hybrid inductive and capacitive AC links for wireless EV charging [20]. The efficiency of the output power can vary for different coupling coil designs. IPT of DC-DC has reached 96% at 200 mm distance of power transfer with an output power of 7 kW, which was conducted in a lab setting. To better understand possible efficiencies of wireless chargers, the efficiency of a commercial static EV charger has achieved 92% for a distance of 610 mm and power level between 3.3 and 10 kW. AC link design is

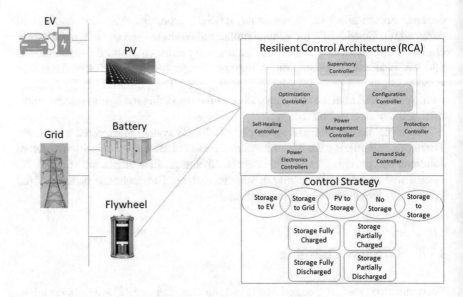

Fig. 15.5 EV Wireless charging and V2V wireless charging in park and during driving

categorized based on power level, coil shape, geometry, frequency, surface power density, air gap, the magnetic field which is generated, magnetic path, or magnetic flux shape. There are options where parallel, horizontal, vertical components are structured to form the charge pads. Plat geometry could be different like rectangular, square, or dish shape.

15.9   V2V Charging

Vehicle-to-vehicle (V2V) charging can effectively support transportation electrification in view of mobility needs while achieving reduced operating costs and impacts on the grid. Figure 15.6 shows a schematic of implementing V2V, where the owner of charging vehicle "A" will receive a request from the charged vehicle "B" to start charging. EV "A" owner will confirm charging start based on SoC and mobility demands and offer price to the owner of EV "B." Once EV "B" owner accepts the price and terms, the charging will start. Both owners will be able to view the charging status. This model could be implemented in more than two EVs where the circuit could be modified to accommodate more charging EVs.

Vehicular ad hoc networks (VANETs) are spontaneous networks based on the concepts of mobile ad hoc networks (MANETs) where a wireless network of mobile devices is established to connect EVs for two-way communications and control with information exchange [1–3].

The network layer management will facilitate the analysis of routing paths among connected vehicles while achieving quality of service and quality of service-optimized link state routing (QoS-OLSR), which is used to find routing paths on multipoint relays (MPRs). These MPRs are utilized to propagate and communicate announcements between the providers and other vehicles in the network.

An effective ad hoc V2V charging protocol will provide connectivity between the providers and consumers within the network. The adaptive network will allow the ad hoc discovery of available providers with exchanged messages. The protocol will also allocate the best provider-consumer pairs for V2V charging in a fully decentralized way with reasonable performance.

Several design options for the receiver and transmitter coils enhance the charging performance with higher efficiencies [16].

There are a number of V2V charging modes depending on charging EV, charged EV, fast or regular, while stopping, or while moving, as shown in Fig. 15.7. It shows a possible onboard charger for possible AC or DC charging options.

Fig. 15.6 V2V implementation

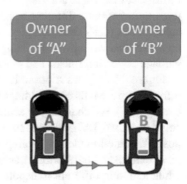

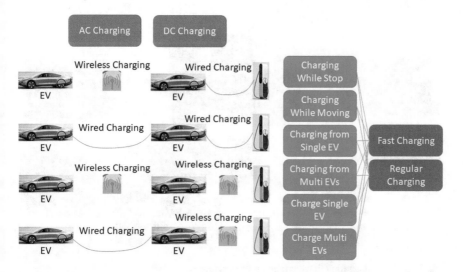

Fig. 15.7 V2V charging options

Management of V2V charging includes several factors, such as energy, economic, mobility, social, emergency, and safety. Energy considerations include SoC and power rates. Economic factors include charging costs. Mobility factors include planned trip distance. Social factors include equity, diversity, and inclusion. Emergency factors include ambulance and VIP priority vehicles. Safety factors include hazard or loss prevention or mitigation.

15.10   Next-Generation Transportation Infrastructure

Transportation infrastructures are evolving quickly because of the technology development of transportation systems as well as growth in mobility demand. Also, there are efforts to transition into smart cities, which directly affect transportation infrastructures. There are a number of initiatives to change transportation infrastructures to incorporate different business models such as shared rides, connected and autonomous (CAV) technologies, and drones. Different road systems and community activities will also have changes, such as emergency vehicles, police vehicles, garbage collection trucks, and other service vehicles (e.g., snow removal trucks).

Figure 15.8 shows a possible demonstration of next-generation transportation infrastructure. Malls and buildings can include a possible swap of battery, charge of battery (BT), EV charging, or share of EV or CAV. Post office and dispatch centers could use CAV, EV, or drones to ship mail. Houses can also use EV and CAV, with possible shared models. Charging stations can charge EV and CAV or offer swapping batteries. The flexible transportation infrastructure will facilitate mobility while adopting different transportation business models to share a ride, share battery, share vehicle ownership, or use CAV.

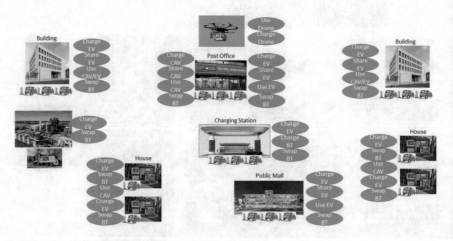

Fig. 15.8 Next-generation transportation infrastructure

15.11   Summary

This chapter provided an analysis of a number of advanced charging technologies, which are explained with basic fundamentals and examples. The development of charging infrastructures reflected potential implementations of transportation electrification infrastructures. Charging techniques and technologies are illustrated and analyzed using different charging models. Vehicle to grid (V2G) is explained with benefits to the grid while supporting mobility demands. Wireless charging techniques are explained where possible charging models and structures are illustrated with related parameters. Vehicle-to-vehicle (V2V) charging approaches are discussed where different models are explained to demonstrate implementation strategies with the associated performance measures. Finally, potential views of fast-charging infrastructures are presented with possible implementation strategies to enable fast charging in different city facilities.

References

1. Q. Zhang, H. Li, L. Zhu, P.E. Campana, H. Lu, F. Wallin, Q. Sun, Factors influencing the economics of public charging infrastructures for EV – A review. Renew. Sust. Energ. Rev. **94**, 500–509 (2018)
2. C.K. Toh, *Ad Hoc Mobile Wireless Networks: Protocols and Systems* (Prentice Hall, 2001). ISBN 9780132442046
3. H. Abualola, H. Otrok, R. Mizouni, S. Singh, A V2V charging allocation protocol for electric vehicles in VANET. Veh. Commun. **33**, 100427 (2022)
4. S. Taghizadeh, M.J. Hossain, N. Poursafar, J. Lu, G. Konstantinou, Multifunctional single-phase EV on-board charger with a new V2V charging assistance capability. IEEE Access **8**, 116812–116823 (2020)
5. M. Wang, Shen, X. (Xuemin), R. Zhang, *Mobile Electric Vehicles: Online Charging and Discharging, Book* (Springer, Cham, 2016)
6. A.M.J. Skulimowski, Z. Sheng, S. Khemiri-Kallel, C. Cérin, C.-H. Hsu, Internet of vehicles, in *Technologies and Services Towards Smart City 5th International Conference, IOV 2018, Paris, France, November 20–22, 2018, Proceedings*, (2018)
7. E. Ucer, R. Buckreus, M.E. Haque, M. Kisacikoglu, Y. Sozer, S. Harasis, M. Guven, L. Giubbolini, Analysis, design, and comparison of V2V chargers for flexible grid integration. IEEE Trans. Ind. Appl. **57**(4), 4143–4154 (2021)
8. M. Shurrab, S. Singh, H. Otrok, R. Mizouni, V. Khadkikar, H. Zeineldin, A stable matching game for V2V energy sharing-a user satisfaction framework. IEEE Trans. Intell. Transp. Syst., 1–13 (2021)
9. R. Zhang, X. Cheng, L. Yang, Flexible energy management protocol for cooperative EV-to-EV charging. IEEE Trans. Intell. Transp. Syst. **20**(1), 172–184 (2019)
10. J. Zhang, Q. Tang, S. Liu, Y. Cao, W. Zhao, T. Liu, S. Xie, Deadline-based V2V charging under spatial resource constraints, in *2021 International Conference on Control, Automation and Information Sciences (ICCAIS)*, (2021), pp. 240–245
11. P. Barbecho Bautista, L. Lemus Cárdenas, L. Urquiza Aguiar, M. Aguilar Igartua, A traffic-aware electric vehicle charging management system for smart cities. Veh. Commun. **20**, 100188 (2019). https://doi.org/10.1016/j.vehcom.2019.100188

12. V. del Razo, H. Jacobsen, Smart charging schedules for highway travel with electric vehicles. IEEE Trans. Transp. Electrif. **2**(2), 160–173 (2016)
13. I. Wagner, Global plug-in electric light vehicle sales from 2015 to 2019. https://www.statista.com/statistics/665774/global-sales-of-plug-in-lightvehicles/ (2020, March 31)
14. IEA, *Global EV Outlook 2020* (IEA, Paris, 2020). https://www.iea.org/reports/global-ev-outlook-2020
15. H.A. Gabbar, A.M. Othman, F. Pino, M. Repetto, Improved performance of flywheel fast charging system (FFCS) using enhanced artificial immune system (EAIS). IEEE Syst. J. **14**(1) (2020)
16. X. Mou, D.T. Gladwin, R. Zhao, H. Sun, Z. Yang, Coil design for wireless vehicle-to-vehicle charging systems. IEEE Access **8**, 172723–172733 (2020)
17. Plugless Power. https://www.pluglesspower.com/
18. When Can We Expect Wireless Charging for Electric Vehicles? [Online]. Available https://www.fleetcarma.com. Accessed 5 Jan 2021
19. M. Wang, R. Zhang, Shen, X. (Sherman), Coordinated V2V fast charging for mobile GEVs based on price control, in *Book: Mobile Electric Vehicles*, (2015), pp. 55–69
20. D. Vincent, P.S. Huynh, N.A. Azeez, L. Patnaik, S.S. Williamson, Evolution of hybrid inductive and capacitive AC links for wireless EV charging – A comparative overview. IEEE Trans. Transp. Electr. **5**(4), 1060–1077 (2019)

Chapter 16
Nuclear-Renewable Hybrid Energy Systems with Charging Stations for Transportation Electrification

**Hossam A. Gabbar, Muhammad R. Abdussami,
Md. Ibrahim Ibrahim Adham, and Ajibola Adeleke**

16.1 Introduction

The plan toward 2030 Sustainable Development included a number of Sustainable Development Goals (SDGs), which provided valuable guidelines to achieve sustainability. The SDGs included important targets such as "affordable and clean energy," which indicated the need to use performance measures of energy systems, including lifecycle costing with reduced GHG (greenhouse gas) emission throughout the lifetime of energy systems [1–2]. Energy supply and demand include electricity, thermal, and fuel, linked to infrastructures, transportation, water, waste, and community services. There are increases in energy demand with higher environmental requirements, which triggered the needs of analyzing different energy technologies including nuclear and renewable energy systems [3–4], in particular with the environmental stresses from fossil fuel-based resources and technologies.

The analysis of integrated nuclear-renewable hybrid energy systems N-R HES showed the clear potential of adopting in different scales. The design of effective N-R HES is based on dynamic modeling and simulation to analyze design and configuration scenarios and potential operation and control strategies so that lifecycle

The original version of this chapter was revised. The correction to this chapter is available at https://doi.org/10.1007/978-3-031-09500-9_18

H. A. Gabbar (✉)
Faculty of Energy Systems and Nuclear Science, Ontario Tech University (UOIT),
Oshawa, ON, Canada

Faculty of Engineering and Applied Science, Ontario Tech University,
Oshawa, ON, Canada
e-mail: Hossam.gabbar@uoit.ca

M. R. Abdussami · Md. I. Adham · A. Adeleke
Faculty of Energy Systems and Nuclear Science, Ontario Tech University (UOIT),
Oshawa, ON, Canada

costing could be analyzed accurately and compared with other traditional scenarios [5]. The N-R HES could be viewed as a hybrid energy system that includes a nuclear reactor that typically produces thermal energy, a turbine to generate electricity from thermal energy, renewable energy technologies, and the expected outcomes linked to industrial processes and applications. There are a number of possible ways to integrate and couple nuclear and renewable energy categorized into three possible categories: tightly coupled HES, thermally coupled HES, and loosely coupled electricity-only HES. Based on several case studies and analyses using simulation models, it is concluded that N-R HES can offer reduced GHG emissions, resilient energy grids, and low cost of energy (COE).

There are a number of potential features of integrating electricity, thermal, and fuel, including chemicals and hydrogen processes. The analysis showed potential benefits toward long-term energy supply and infrastructure planning compared to fossil fuels. The International Atomic Energy Agency (IAEA) is leading in developing a technical framework and introducing guidelines to support effective deployments of N-R HES in different scales [6–8]. The studies compared possible scenarios with a number of combinations of nuclear power plants, wind energy, hydrogen generation facility, gas generators, combined heat and power (CHP) systems, and other energy resources and technologies. The optimization and cost analysis reflected potential performance assessment in view of energy demand and energy markets, considering depreciation rate, discount rate, timeline, internal rate of return (IRR), levelized cost of energy (LCOE), net present value (NPV), and payback period. The study included different scales of nuclear power plants, such as small-scale nuclear reactors, small modular reactors (SMRs), and micro modular reactors (MMRs).

In addition, U-battery is expected to demonstrate this type of reactor by 2026 [9–12]. Very small reactors are also developed, for example, eVinci™ micro reactor with combined heat and power (CHP) rated from 200 kWe to 5 MWe [13]. The studies on small-scale, nuclear-renewable micro hybrid energy systems (N-R MHES) showed potential applications toward resilient energy solutions. One key application of N-R HES is transportation electrification, whereby it can be used for charging infrastructures as explored in [14]. Hybrid energy storage-based fast-charging station is integrated with micro modular reactor to charge electric vehicles due to its small size, reliability, and security. Understanding the integration mechanisms using modeling and simulation is required, which will be explained in this chapter.

16.2 System Description

In this case study, the selected location is Ontario Tech University (UOIT), North Campus, Oshawa, Canada. This example evaluates the economic benefits of nuclear-renewable integrated systems. The *microgrid* of the specified location consists of a wind turbine (WT), solar photovoltaic (PV), SMR, diesel generator, electric load, and power electronic devices. Diesel generator is introduced as a backup during the unavailability of SMR. The SMR requires refueling at 10-year intervals. During the refueling period, alternative generation sources, such as diesel generators, are

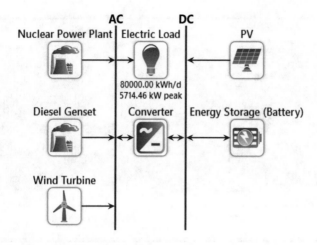

Fig. 16.1 Case study on microgrid with SMR (system schematic)

Table 16.1 Optimal configuration of the system

Parameters	Value
NPC ($)	118 million
COE ($/kWh)	0.246
No. of PV panels	3208
No. of WT	383
No. of SMR	2
Capacity of the diesel generator (MW)	6.3
Capacity of the battery bank (MWh)	5.2
Capacity of the converter (MW)	3.9

needed to support the demand. Figure 16.1 represents the off-grid system configuration of the microgrid [1]. In this case study, the integration between the proposed microgrid and SMR is planned on the grid side, which is classified as a loosely coupled method. Detailed specifications of the system component are presented in Table 16.1.

The load profile of the selected data is represented in Fig. 16.2, as per reference [1]. The electric load data were collected for Ontario Tech University, North Campus, Oshawa, Ontario, Canada, in 2018. The data have been collected from the different facilities of Ontario Tech University.

The study intends to determine the optimal configuration of the hybrid system. Net present cost (NPC) is considered the optimization problem's objective function. HOMER Pro software is used in this study for optimization. HOMER Pro is one of the most popular optimization tools for hybrid energy systems. It simulates all possible feasible system configurations and orders based on NPC. The decision variables of the integrated system are represented as follows:

$$0 \le N_{PV} \le N_{PV}^{\max} \qquad (16.1)$$

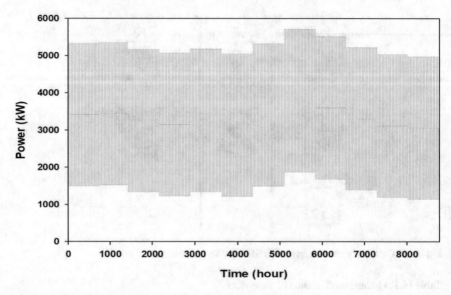

Fig. 16.2 Total electric load served by the system in 2018

$$0 \leq N_{WT} \leq N_{WT}^{max} \tag{16.2}$$

$$0 \leq N_{SMR} \leq N_{SMR}^{max} \tag{16.3}$$

$$0 \leq Cap_{diesel} \leq Cap_{diesel}^{max} \tag{16.4}$$

$$0 \leq Cap_{battery} \leq Cap_{battery}^{max} \tag{16.5}$$

$$0 \leq Cap_{converter} \leq Cap_{converter}^{max} \tag{16.6}$$

16.3 Case Study

In the simulation, the "cycle charging (CC)" strategy is applied. In the CC control algorithm, the RESs and SMR will initially serve the electricity demand. After accomplishing the electricity demand, the surplus (if any) electricity will be consumed by battery banks for charging. Conversely, battery banks discharge energy if the renewables and SMR cannot meet the demand. In this investigation, the SMR operates continuously at its rated power irrespective of the load demand since the "load-following (LF)" control strategy is not cost-effective for NPP operation. The project lifetime is considered 60 years. The optimal configuration of the hybrid energy system is presented in Table 16.1.

Fig. 16.3 Energy production share by different generation sources

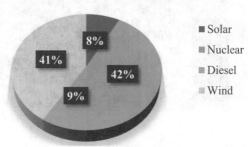

Contribution in total energy prodution by different sources

- ■ Solar
- ■ Nuclear
- ■ Diesel
- ▪ Wind

8%

41%

42%

9%

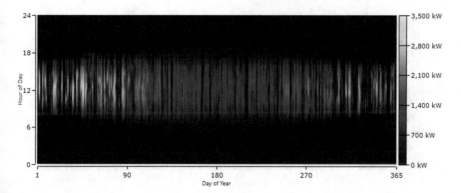

Fig. 16.4 PV panel output

Figure 16.3 shows the contribution of different generation sources in energy production. Due to reasonable wind speed in the project location, a considerable percentage of wind energy contributes to the total energy generation. Solar PV panels generate a small portion (8%) of the total energy because of unfavorable solar irradiance and lower PV panel efficiency. Diesel Genset contributes by a small part (9%) as it operates only when the renewables and SMR cannot fulfill the load demand.

16.4   Results

This section describes the results obtained from the simulation, which shows simulation outcomes for the different design options. Figures 16.4, 16.5, 16.6, and 16.7 present the total PV panel output, WT output, diesel generator output, and SoC (state of charge) of the battery bank, respectively.

Since the performance of a nuclear-renewable hybrid energy system depends on several factors, two sensitivity analyses have been conducted in this study to assess

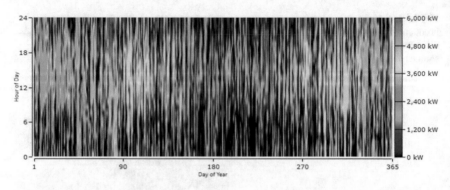

Fig. 16.5 Wind turbine output

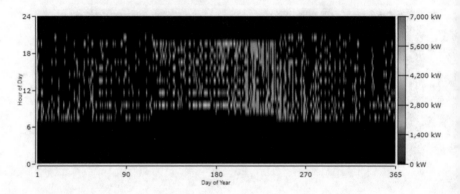

Fig. 16.6 Diesel generator output

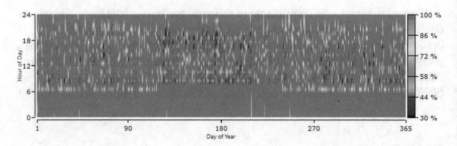

Fig. 16.7 SoC of the battery bank

the impact of solar irradiance, wind speed, and variation of load demand on system NPC.

Figure 16.8 represents the impact on NPC and optimal system configuration due to variation in system demand. In this sensitivity analysis, the average electric demand is varied by ±40% and ±20%. The sensitivity analysis results show that nuclear-renewable integration is always an economical choice regardless of system demand variation. All the cases include nuclear reactors, PV panels, and WT.

Sensitivity		Architecture						
Electric Load Scaled Average (kWh/d)	PV (kW)	Wind Turbine	Nuclear Power Plant (kW)	Diesel Genset (kW)	Energy Storage (Battery)	Converter (kW)	Dispatch
48,000	1,435	54	3,000		2,320	839	CC
64,000	3,201	251	2,000	8,900	4,243	3,046	CC
80,000	2,850	392	2,000	8,900	5,399	4,096	CC
96,000	4,222	500	2,000	8,900	6,908	5,186	CC
112,000	4,379	495	3,000	8,900	6,984	5,283	CC

Fig. 16.8 Sensitivity of NPC due to variation in average demand

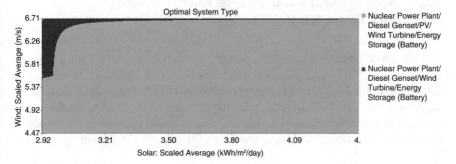

Fig. 16.9 Sensitivity of NPC due to variation in solar irradiance and wind speed (graphical form)

Sensitivity		Architecture						
Solar Scaled Average (kWh/m²/day)	Wind Scaled Average (m/s)	PV (kW)	Wind Turbine	Nuclear Power Plant (kW)	Diesel Genset (kW)	Energy Storage (Battery)	Converter (kW)	Dispatch
2.92	4.47	4,912	242	3,000	6,300	4,322	3,305	CC
2.92	5.59		462	2,000	6,300	4,044	2,924	CC
2.92	6.71		500	1,000	6,300	4,896	3,572	CC
3.65	4.47	4,896	182	3,000	6,300	4,375	3,304	CC
3.65	5.59	3,201	362	2,000	6,300	5,104	3,898	CC
3.65	6.71		500	1,000	6,300	4,896	3,572	CC
4.38	4.47	4,969	171	3,000	6,300	4,605	3,301	CC
4.38	5.59	4,178	328	2,000	6,300	5,097	3,931	CC
4.38	6.71		497	1,000	6,300	4,896	3,527	CC

Fig. 16.10 Sensitivity of NPC due to variation in solar irradiance and wind speed (tabular form)

Besides, the optimization result does not recommend either a stand-alone renewable-based energy system or a stand-alone nuclear system.

In the second sensitivity analysis, the solar irradiance and the wind speed are varied by ±20% to determine the changes in optimal system configuration. The average solar irradiance and wind speed for the base case were 3.65 kWh/m²/day and 5.59 m/s, respectively. Figure 16.9 shows the result of the second sensitivity analysis. Figure 16.9 tells that nuclear/diesel/PV/WT/battery-based energy system can serve the demand for a significant range despite the variation in solar irradiance and wind speed (marked in light green). However, the optimization does not suggest installing solar panels if the wind speed is relatively high (marked in dark blue). The increased wind speed allows more WT to be included within the system, which eliminates the necessity of the solar PV panel indicated in Fig. 16.10. Though Figs. 16.9 and 16.10

offer the same information, Fig. 16.10 represents the detailed configuration of differ-ent optimal systems based on solar irradiance and wind speed variation. Also, Fig. 16.10 indicates that it is required to include nuclear with renewables in all cases. The optimal configurations for different scenarios caused by resource data variation (e.g., solar irradiance and wind speed) are present in Fig. 16.10.

16.5   Nuclear-Renewable Hybrid Energy Systems with Fast-Charging Station

The *microgrid* block diagram above consists of a wind turbine (WT), solar photo-voltaic (PV), SMR, diesel generator, electric load, power electronic devices, and a fast-charging station. Diesel generator is introduced as a backup during the unavail-ability of SMR. The SMR requires refueling at 10-year intervals. During the refuel-ing period, alternative generation sources, such as diesel generators, are needed to support the demand. Figure 16.11 represents the off-grid system configuration of

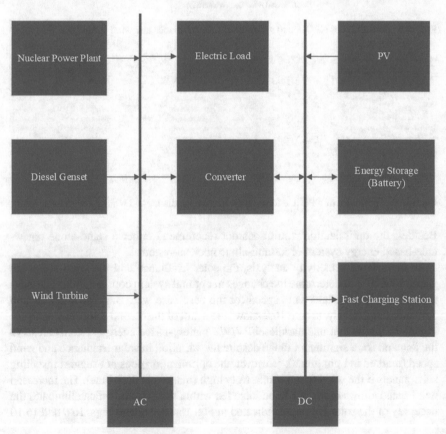

Fig. 16.11 Block diagram of the N-R hybrid energy system integrated with fast-charging station

the microgrid [1]. Detailed specifications of the fast-charging station are presented in Table 16.2.

16.6   Fast-Charging Station Design

The block diagram below is a Simulink model for the fast-charging station with an induction motor used as the flywheel and the battery as the energy storage system (Fig. 16.12).

Figure 16.13 shows the flywheel rotor speed within the fast-charging station. The DC bus voltage of the fast-charging station model is shown in Fig. 16.14.

16.6.1   Charging Mode

During the charging mode, the state-of-charge output is shown in Fig. 16.15.

Table 16.2   Parameter used for the simulation

C_{DC}	2.2 mF	L_{line}	3.8 mH
R_{line}	0.24 Ω	V_{grid}(P-P)	325 V
L_0	10.46 mH	L_s	10.76 mH
L_r	10.76 mH	R_s	0.0148 Ω
R_r	0.0093 Ω	J	10 kgm^2
P_{IM}	1.5 kW	p	2

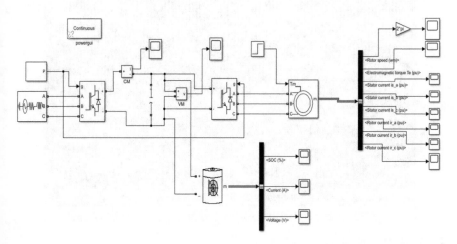

Fig. 16.12   Model block for the flywheel-based fast-charging station

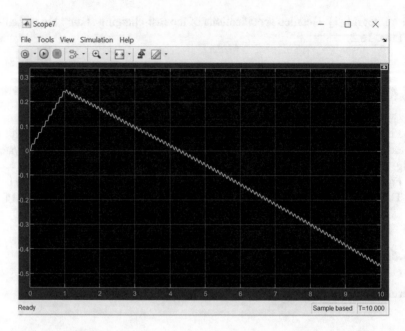

Fig. 16.13 Flywheel rotor speed

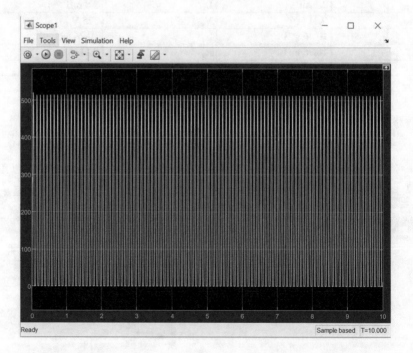

Fig. 16.14 DC bus voltage

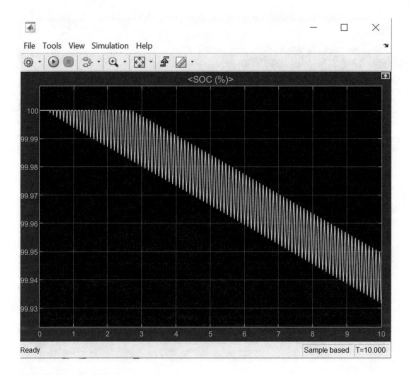

Fig. 16.15 Battery SoC during charging

16.6.2 Discharging Mode

When the battery is connected and discharging, the output of the system is shown in Fig. 16.16.

16.7 Summary

The analysis reveals that nuclear-renewable integrated systems have significant economic benefits. The sensitivity analyses strengthen the results obtained in the base case. The sensitivity analyses demonstrate the superiority of nuclear-renewable integrated systems irrespective of the variation in system demand, solar irradiance, and wind speed. The main benefit of integrating nuclear energy with renewable energy sources to power the fast-charging station is the intermittency of the RES. The renewable sources will not be able to power the station at all times, especially at night for the PV, and when the weather is not windy enough to provide the wind turbine with enough energy. The nuclear energy source will always be available to power the fast-charging station making the system more resilient.

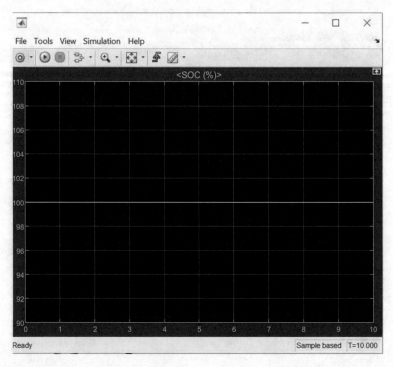

Fig. 16.16 Battery SoC during discharging

Acknowledgment Authors would like to thank members in the Smart Energy Systems Laboratory (SESL), in particular Dr. Yasser Elsayed for his support to the work.

References

1. H.A. Gabbar, M.R. Abdussami, Feasibility analysis of grid-connected nuclear-renewable micro hybrid energy system, in *2019 IEEE 7th International Conference on Smart Energy Grid Engineering (SEGE)*, 2019, pp. 294–298
2. H.A. Gabbar, M.R. Abdussami, M.I. Adham, Optimal Planning of Nuclear-Renewable Micro-Hybrid Energy System by Particle Swarm Optimization, in *IEEE Access*, 2020, pp. 1–1. https://doi.org/10.1109/ACCESS.2020.3027524
3. Cost competitiveness of micro-reactors for remote markets, in *Nuclear Energy Institute*. https://www.nei.org/resources/reports-briefs/cost-competitiveness-micro-reactors-remote-markets. Accessed 11 June 2020
4. Nuclear Power Economics | Nuclear Energy Costs – World Nuclear Association. https://www.world-nuclear.org/information-library/economic-aspects/economics-of-nuclear-power.aspx. Accessed 20 Mar 2020
5. N.E.B. Government of Canada, NEB – Market Snapshot: The cost to install wind and solar power in Canada is projected to significantly fall over the long term, Feb. 13, 2020. https://www.cer-rec.gc.ca/nrg/ntgrtd/mrkt/snpsht/2018/11-03cstnstllwnd-eng.html. Accessed 24 July 2020

6. Solar Panel Maintenance Costs I Solar Power Maintenance Estimates, *Fixr.com*. https://www. fixr.com/costs/solar-panel-maintenance. Accessed 16 Apr 2020
7. What Is the Lifespan of a Solar Panel? https://www.engineering.com/DesignerEdge/ DesignerEdgeArticles/ArticleID/7475/What-Is-the-Lifespan-of-a-Solar-Panel.aspx. Accessed 16 Apr 2020
8. US wind O&M costs estimated at $48,000/MW; Falling costs create new industrial uses: IEA I New Energy Update. https://analysis.newenergyupdate.com/wind-energy-update/us-wind-om-costs-estimated-48000mw-falling-costs-create-new-industrial-uses-iea. Accessed 17 Apr 2020
9. T. Stehly, P. Beiter, D. Heimiller, G. Scott, 2017 cost of wind energy review. *Renewable Energy*, 61 (2018)
10. S.J. Ericson, D.R. Olis, A Comparison of Fuel Choice for Backup Generators, NREL/ TP--6A50-72509, 1505554, Mar 2019. https://doi.org/10.2172/1505554
11. Canada diesel prices, 18-Nov-2019 I GlobalPetrolPrices.com. https://www.globalpetrolprices. com/Canada/diesel_prices/. Accessed 25 Nov 2019
12. S. Kharel, B. Shabani, Hydrogen as a long-term large-scale energy storage solution to support renewables. *Energies* **11**(10), 2825 (2018). https://doi.org/10.3390/en11102825
13. S.C. Johnson, et al., Selecting favorable energy storage technologies for nuclear power, in *Storage and Hybridization of Nuclear Energy* (Elsevier, 2019) pp. 119–175
14. M.R. Abdussami, H.A. Gabbar, Nuclear-powered hybrid energy storage-based fast charging station for electrification transportation, in *2019 IEEE 7th International Conference on Smart Energy Grid Engineering (SEGE)*, 2019, pp. 304–308. https://doi.org/10.1109/ SEGE.2019.8859878

Chapter 17
Transactive Energy for Charging Infrastructures

Hossam A. Gabbar

Acronyms

EV	Electric vehicle
EB	Electric bus
ET	Electric truck
FC	Fuel cell
FCS	Fast-charging station
MEG	Micro energy grid
FCV	Fast-charging stations
GHG	Greenhouse gas
TC	Transactive charging
TE	Transactive energy
WT	Wind turbine

Nomenclature

c	Charging unit index
i	Vehicle index
s	Charging station index
t	Time [h]
SOC^{EV}_{min}	Minimum state of charge of EV battery [%]

This chapter is contributed by Hossam A. Gabbar.

H. A. Gabbar (✉)
Energy Systems and Nuclear Science, University of Ontario Institute of Technology,
Oshawa, ON, Canada
e-mail: Hossam.gabbar@uoit.ca

© Springer Nature Switzerland AG 2022
H. A. Gabbar, *Fast Charging and Resilient Transportation Infrastructures in
Smart Cities*, https://doi.org/10.1007/978-3-031-09500-9_17

SOC_i^{EV}	SOC of EV i
SOC_{max}^{EV}	Maximum state of charge of EV battery [%]
SOC_{R1KM}^{EV}	SOC required for EV to travel for 1 km distance [%]
$DTS_{s,i}^{EV}$	Distance to charging station s for EV i [km]
BC_i^{EV}	Battery capacity of EV i [kWh]
$PT_{c,i,t}^{EV}$	Power transfer from/to EV i, at time t and at charging unit c [kW]
$PT_{c,i,t}^{CtG}$	Power transfer from charging unit (c) to grid at time t [kW]
$PT_{c,i,t}^{GtC}$	Power transfer from grid to charging unit (c) at time t [kW]
$PT_{c,i,t}^{CtBT}$	Power transfer from charging unit (c) to battery at time t [kW]
$PT_{c,i,t}^{BTtC}$	Power transfer from battery to charging unit (c) at time t [kW]
$PT_{c,i,t}^{CtUC}$	Power transfer from charging unit (c) to ultra-capacitor at time t [kW]
$PT_{c,i,t}^{UCtC}$	Power transfer from ultra-capacitor to charging unit (c) at time t [kW]
$PT_{c,i,t}^{CtFW}$	Power transfer from charging unit (c) to flywheel at time t [kW]
$PT_{c,i,t}^{FWtC}$	Power transfer from flywheel to charging unit (c) at time t [kW]
$PT_{c,i,t}^{PVtC}$	Power transfer from PV to charging unit (c) at time t [kW]
$PT_{c,i,t}^{FCtC}$	Power transfer from FC to charging unit (c) at time t [kW]
$PT_{c,i,t}^{NRtC}$	Power transfer from NR to charging unit (c) at time t [kW]
$PT_{c,i,t}^{WTtC}$	Power transfer from WT to charging unit (c) at time t [kW]
$PT_{c,i,t}^{PVtBT}$	Power transfer from PV to battery at time t [kW]
$PT_{c,i,t}^{PVtUC}$	Power transfer from PV to ultra-capacitor at time t [kW]
$PT_{c,i,t}^{PVtFW}$	Power transfer from PV to flywheel at time t [kW]
$PT_{c,i,t}^{FCtBT}$	Power transfer from FC to battery at time t [kW]
$PT_{c,i,t}^{FCtUC}$	Power transfer from FC to ultra-capacitor at time t [kW]
$PT_{c,i,t}^{FCtFW}$	Power transfer from FC to flywheel at time t [kW]
$PT_{c,i,t}^{NRtBT}$	Power transfer from NR to battery at time t [kW]
$PT_{c,i,t}^{NRtUC}$	Power transfer from NR to ultra-capacitor at time t [kW]
$PT_{c,i,t}^{NRtFW}$	Power transfer from NR to flywheel at time t [kW]
$PT_{c,i,t}^{WTtBT}$	Power transfer from WT to battery at time t [kW]
$PT_{c,i,t}^{WTtUC}$	Power transfer from WT to ultra-capacitor at time t [kW]
$PT_{c,i,t}^{WTtFW}$	Power transfer from WT to flywheel at time t [kW]
$En_{c,i,t}^{GTBT}$	Energy supplied from the grid to the battery on charging EV i, at station c and time t [kW]

$En_{c,i,t}^{GTUC}$	Energy supplied from the grid to the UC on charging EV i, at station c and time t [kW]
$En_{c,i,t}^{GTFW}$	Energy supplied from the grid to the FW on charging EV i, at station c and time t [kW]
$En_{c,i,t}^{GTC}$	Energy supplied from the grid to the charging station c on charging EV i and time t [kW]
$En_{c,i,t}^{GTNR}$	Energy supplied from the grid to the NR on charging EV i, at station c and time t [kW]
$En_{c,i,t}^{STG}$	Energy supplied from station c to the grid on charging EV i and time t [kW]
$En_{c,i,t}^{BTTG}$	Energy supplied from battery to the grid at station c on charging EV i and time t [kW]
$En_{c,i,t}^{UCTG}$	Energy supplied from UC to the grid at station c on charging EV i and time t [kW]
$En_{c,i,t}^{FWTG}$	Energy supplied from flywheel to the grid at station c on charging EV i and time t [kW]
$En_{c,i,t}^{CTG}$	Energy supplied from station to the grid at station c on charging EV i and time t [kW]
$En_{c,i,t}^{NRTG}$	Energy supplied from NR to the grid at station c on charging EV i and time t [kW]
$C_{PV,t}^{LCOE}$	PV levelized cost of energy (LCPV) at time t [$/kWh]
$C_{WT,t}^{LCOE}$	Wind turbine levelized cost of energy (LCWT) at time t [$/kWh]
$C_{FC,t}^{LCOE}$	Fuel cell levelized cost of energy (LCFC) at time t [$/kWh]
$C_{NR,t}^{LCOE}$	Nuclear reactor levelized cost of energy (LCNR) at time t [$/kWh]
$C_{Grid,t}^{COE}$	Grid cost of energy (COEG) at time t [$/kWh]
$C_{BT,t}^{LCOE}$	Levelized energy cost of FCS battery (LCBT) at time t [$/kWh]
$C_{UC,t}^{LCOE}$	Levelized energy cost of FCS ultra-capacitor (LCUC) at time t [$/kWh]
$C_{FW,t}^{LCOE}$	Levelized energy cost of FCS flywheel (LCFW) at time t [$/kWh]
$EP_{BT,c,i,t}^{Chg}$	Energy price of charging FCS battery for EV i, at charging unit c and at time t [$/kWh]
$EP_{BT,c,i,t}^{Dis}$	Energy price of discharging FCS battery for EV i, at charging unit c and at time t [$/kWh]
$EP_{UC,c,i,t}^{Chg}$	Energy price of charging FCS UC for EV i, at charging unit c and at time t [$/kWh]
$EP_{UC,c,i,t}^{Dis}$	Energy price of discharging FCS UC for EV i, at charging unit c and at time t [$/kWh]
$EP_{FW,c,i,t}^{Chg}$	Energy price of charging FCS FW for EV i, at charging unit c and at time t [$/kWh]
$EP_{UC,c,i,t}^{Dis}$	Energy price of discharging FCS FW for EV i, at charging unit c and at time t [$/kWh]

$EP_{EV,c,i,t}^{Chg}$	Energy price of charging EV i, at charging unit c and at Time t [\$/kWh]
$EP_{EV,c,i,t}^{Dis}$	Energy price of discharging EV i (V2G) at charging unit c and at time t [\$/kWh]
$EP_{PV,c,i,t}^{Chg}$	Energy price of PV for charging EV i, at charging unit c and at time t [\$/kWh]
$EP_{FC,c,i,t}^{Chg}$	Energy price of FC for charging EV i, at charging unit c and at time t [\$/kWh]
$EP_{WT,c,i,t}^{Chg}$	Energy price of WT for charging EV i, at charging unit c and at time t [\$/kWh]
$EP_{NR,c,i,t}^{Chg}$	Energy price of NR for charging EV i, at charging unit c and at time t [\$/kWh]
$EP_{CU,c,i,t}^{min,Chg}$	Min energy bidding price for EV i, at charging unit c and at time t [\$/kWh]
$EP_{CU,c,i,t}^{max,Chg}$	Max energy bidding price for EV i, at charging unit c and at time t [\$/kWh]
$EP_{CU,c,i,t}^{min,V2G}$	Min energy bidding price for V2G of EV i, at charging unit c and at time t [\$/kWh]
$EP_{CU,c,i,t}^{max,V2G}$	Max energy bidding price for V2G of EV i, at charging unit c and at time t [\$/kWh]
$EP_{EV,c,i,t}^{min,Chg}$	Min energy bidding price for charging of EV i, at time t [\$/kWh]
$EP_{EV,c,i,t}^{max,Chg}$	Max energy bidding price for charging of EV i, at time t [\$/kWh]
$EP_{EV,c,i,t}^{min,V2G}$	Min energy bidding price for V2G of EV i, at time t [\$/kWh]
$EP_{EV,c,i,t}^{max,V2G}$	Max energy bidding price for V2G of EV i, at time t [\$/kWh]
$EP_{BT,c,i,t}^{min,Chg}$	Min energy bidding price for charging FCS battery for EV i, at time t [\$/kWh]
$EP_{BT,c,i,t}^{max,Chg}$	Max energy bidding price for charging FCS battery for EV i, at time t [\$/kWh]
$EP_{BT,c,i,t}^{min,Dis}$	Min energy bidding price for discharging FCS battery of EV i, at time t [\$/kWh]
$EP_{BT,c,i,t}^{max,Dis}$	Max energy bidding price for discharging FCS battery of EV i, at time t [\$/kWh]
$EP_{UC,c,i,t}^{min,Chg}$	Min energy bidding price for charging FCS UC for EV i, at time t [\$/kWh]
$EP_{UC,c,i,t}^{max,Chg}$	Max energy bidding price for charging FCS UC for EV i, at time t [\$/kWh]
$EP_{UC,c,i,t}^{min,Dis}$	Min energy bidding price for discharging FCS UC of EV i, at time t [\$/kWh]
$EP_{UC,c,i,t}^{max,Dis}$	Max energy bidding price for discharging FCS UC of EV i, at time t [\$/kWh]
$EP_{FW,c,i,t}^{min,Chg}$	Min energy bidding price for charging FCS FW for EV i, at time t [\$/kWh]
$EP_{FW,c,i,t}^{max,Chg}$	Max energy bidding price for charging FCS FW for EV i, at time t [\$/kWh]

$EP_{FW,c,i,t}^{min,Dis}$ Min energy bidding price for discharging FCS FW of EV i, at time t [$/kWh]

$EP_{FW,c,i,t}^{max,Dis}$ Max energy bidding price for discharging FCS FW of EV i, at time t [$/kWh]

$EP_{PV,c,i,t}^{min}$ Min energy bidding price of PV for EV i, at time t [$/kWh]

$EP_{PV,c,i,t}^{max}$ Max energy bidding price of PV for EV i, at time t [$/kWh]

$EP_{FC,c,i,t}^{min}$ Min energy bidding price of FC for EV i, at time t [$/kWh]

$EP_{FC,c,i,t}^{max}$ Max energy bidding price of FC for EV i, at time t [$/kWh]

$EP_{NR,c,i,t}^{min}$ Min energy bidding price of NR for EV i, at time t [$/kWh]

$EP_{NR,c,i,t}^{max}$ Max energy bidding price of NR for EV i, at time t [$/kWh]

$EP_{WT,c,i,t}^{min}$ Min energy bidding price of WT for EV i, at time t [$/kWh]

$EP_{WT,c,i,t}^{max}$ Max energy bidding price of WT for EV i, at time t [$/kWh]

17.1 Introduction

Transactive energy (TE) is offering a marketplace for peer-to-peer energy trading and exchange. It is important to identify the specified zone or portion of the energy distribution network that will be served [1]. The transactive energy approach will enable consumers and prosumers to benefit from collaboration. Transactive energy can also support the deployment of microgrids with an evaluation of energy costing in view of internal costs within each microgrid [2]. The evaluation of different scenarios and trade parameters and constraints are evaluated with simulation to examine options prior to actual execution [3]. The selection and evaluation of constraints to couple electricity and gas networks are also discussed to show balancing electricity and gas networks in normal and emergencies [4]. Transactive energy is applied to charging infrastructures with different configurations. Transactive energy is applied on charging of EV in buildings with local PV where cost framework is applied considering energy supply from the grid, local PV energy generation, and local energy storage [5]. Transactive energy is applied on PV-based EV charging in parking lots [6]. Scheduling of EV charging can be optimized based on a collaborative cost model using transactive energy [7]. Energy management should incorporate charging schedules based on cost models from the transactive energy framework [8]. There are a number of applications where energy trade is developed and implemented, such as simultaneous wireless information and power transfer (SWIPT) [9]. The trade mechanism is important to achieve fair transactive energy while reducing CO2 emissions in the energy production and supply chain [10]. Energy trade can be implemented across countries. In Europe, trade of energy between different countries, such as Turkey, is presented to show negotiation mechanisms to ensure satisfied energy supply [11]. Energy is coupled with water networks. The trade can be applied on energy-water nexus, where coupling and value exchange are represented in the trade process [12].

This chapter is discussing framework of transactive energy and its relation with charging infrastructures. Different model parameters are defined and used to evaluate different scenarios of transactive energy and their links with transactive charging and mobility models.

17.2 Transactive Energy for Charging Station

Transactive energy is applied on charging station with local energy resources from renewable energy sources and energy storage. The concept of transactive energy and transactive charging is important to be coupled to ensure good strategies for supplying energy from grid, energy resources, and energy storage. The flow of energy and charging requests are shown in Fig. 17.1. FC-T is the trade agent linked with the trade agent for fuel cell charging and discharging. PV-T is the trade agent for energy supplied by PV. Similarly, WT-T is wind turbine trade, G-T is grid trade, NR-T is nuclear reactor trade, UC-T is ultra-capacitor trade, BT-T is battery trade, and FW-T is flywheel trade.

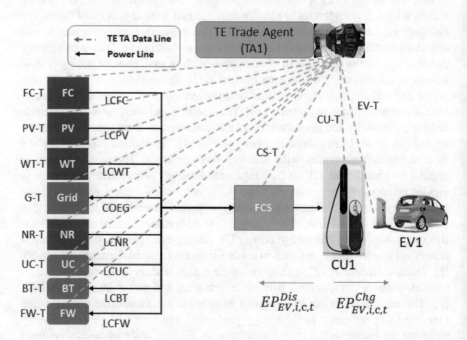

Fig. 17.1 Transactive energy and transactive charging

The levelized energy costs are defined for each component. LCFW is the levelized cost of energy of FW. Similarly, LCBT is the levelized cost of battery. LCUC is the levelized energy cost of the ultra-capacitor. LCNR is the levelized energy cost of the nuclear reactor. COEG is the cost of energy of the grid. LCWT is the levelized energy cost of wind turbine. LCPV is the levelized energy cost of PV. LCFC is the levelized energy cost of fuel cell.

The detailed design of the charging station with integrated trade agent, while considering energy resources and storage to support incoming vehicles, is shown in Fig. 17.2. Renewable energy sources include PV, WT, and FC. Nuclear reactor (NR) is integrated to cover high charging load. Hybrid energy storage system includes battery, ultra-capacitor, and flywheel. The analysis is based on incoming EVs, while EBs and ETs will be handled similarly. The model shows multiple parking spots connected to each charging unit. Controllers are integrated with each component to achieve stable and optimum operation. Trade gent (TA1) is assigned to manage transactive energy in the charging station based on defined parameters, constraints, and optimization objective functions.

Parameters are defined for each incoming EV, including minimum SOC, maximum SOC, and current SOC. Preferred minimum and maximum energy prices are defined for buying from the grid/station and selling to the grid/station. Energy and power price parameters are defined for each energy system component within the station, including minimum and maximum price to buy/sell. Energy transfer among energy sources and storage and the charging unit are defined and used in the management scheme of the transactive energy of the charging station. The energy trade is implemented with direct data communications between all controllers and the

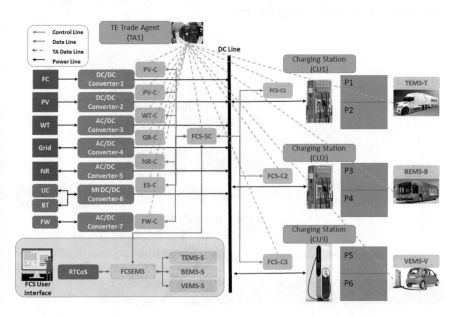

Fig. 17.2 Transactive energy for EV charging station

trade agent and the connection with incoming EVs. Energy trade process will achieve maximum benefits to utilities, EV owner/driver, and charging station owner/operator. This is achieved via trade and bidding with different pricing and constraints with optimization objective functions.

17.2.1   Condition to Start Searching for Charging Station

The mechanism to find the nearest charging station and establish the connection to book charging is shown in Fig. 17.3.

The algorithm is based on finding the list of the nearest charging stations based on EV's current location. The SOC required to reach each charging station is calculated to ensure that EV can reach the charging station based on the current SOC. The proposed algorithm is also seeking the nearest V2V charging units. The detailed algorithm is shown in Table 17.1.

End of charging condition: when SOC_i^{EV} reaches SOC_{max}^{EV}.

Energy levelized cost is calculated for each component based on lifecycle cost, while energy price is the actual energy price for each component. Energy price is estimated based on the trading process with bidding among agents of each component and the incoming EV. Energy price of charging EV i at time t from charging unit c is given by $EP_{EV,c,i,t}^{Chg}$, while the energy price for discharging EV to the grid, i.e., V2G, is $EP_{EV,c,i,t}^{Dis}$. EV can decide to stay longer to charge and then discharge (then charge again) if it can profit with peak and bidding prices while benefiting from the availability of renewable energy sources.

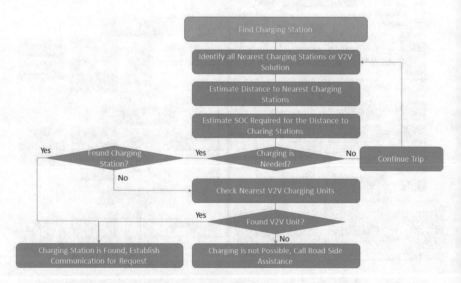

Fig. 17.3 Mechanism to find charging option on the road

Table 17.1 Algorithm to find the nearest charging station

Condition for EV to start seeking charging station:
$\forall s$: for all nearest charging stations
$DTS_{s,i}^{EV} * SOC_{R1KM}^{EV} < SOC_{EV}^{min}$
Identify the nearest charging stations CS(s) based on distance, operating, road conditions
$\forall s$: for all nearest charging stations s (including charging units c within the charging station s)
$\min(EP_{EV,c,i,t}^{Chg})$: Minimum of energy price from all stations to EV i

The increment of SOC per unit time is calculated as $\dfrac{PT_{c,i,t}^{EV}}{BC^{EV}} * \Delta t$. This is used to estimate how long EV should be charged to reach maximum SOC based on power transfer rate from/to EV.

The energy cost calculation is based on how much energy each source provides. The levelized energy costs are considered the reference model to satisfy the lifecycle cost of the energy resource. Levelized energy cost of PV (LCPV) is $C_{PV,t}^{LCOE}$, which is the cost of energy supplied from PV. Similarly, levelized costs from a wind turbine (LCWT), fuel cell (LCFC), and nuclear reactor (LCNR) are calculated similarly. The cost of energy from the grid (COEG) is defined based on the FIT in view of peak pricing. The levelized cost of storing energy in the battery (LCBT) is $C_{BT,t}^{LCOE}$, while the levelized costs of storing energy in the ultra-capacitor (LCUC) and flywheel (LCFW) are $C_{UC,t}^{LCOE}$ and $C_{FW,t}^{LCOE}$, respectively. The calculation of levelized cost is important to include other lifecycle cost elements, such as capital, operation, maintenance, and installation. The cost of energy at the charging unit c is estimated based on the energy flow from grid, PV, wind turbine, fuel cell, nuclear reactor, and energy storage. Power transfer is monitored and used to estimate energy costs. Power is transferred among charging units, PV, wind turbine, nuclear reactor, grid, fuel cell, energy storage, and EV. Power transfer from charging unit c (i.e., EV connected with the charging unit) to the grid at time t is defined as $PT_{c,i,t}^{CtG}$.

17.3 Management of Transactive Charging

The trading mechanism is based on the defined local trade policies, constraints, and strategies. Policies are used to state guidelines for each unit to govern cost estimation, including minimum and maximum with respect to time of the day and in view of charging load profile. Energy price constraints are defined with minimum and maximum for each trade unit, including EV and charging unit. Minimum energy price allowed in the trade for V2G at charging unit c is defined as $EP_{EV,c,i,t}^{min,V2G}$. Similarly, maximum energy price allowed for the trade for V2G at the charging station is $EP_{EV,c,i,t}^{max,V2G}$. The trade limits are also defined for charging unit to EV. Similarly, trade limits are defined for EV for both energy flow to EV and from EV to the station unit. Each trade agent will negotiate price based on local benefits using AI optimization

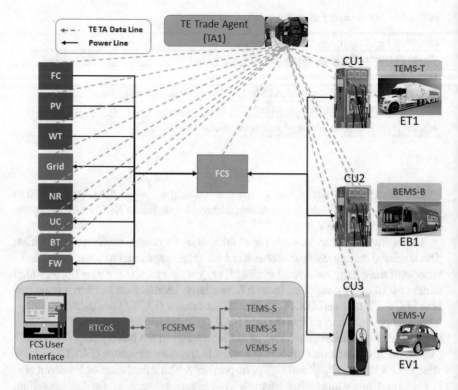

Fig. 17.4 Energy flow of charging station with transactive energy

algorithms to ensure the best local and global benefits. Figure 17.4 shows energy flow between components and charging units, with trade communication data lines and power flow lines.

The framework for transactive energy and transactive charging is shown in Fig. 17.5.

The trade will maximize profit for charging station and grid and minimize cost of energy for EV charging. The three-way optimization model is developed to ensure each view's local and global optimum values.

17.3.1 Grid

In order to estimate the profit to the grid (utilities), the following parameters are defined:

Total energy supplied from the grid to the stations is estimated based on total energy supplied to battery, ultra-capacitor, flywheel, and charging units (minus V2G), as shown in Eq. (17.1):

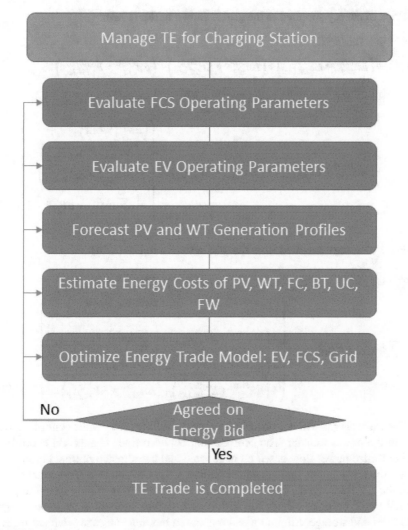

Fig. 17.5 Transactive energy framework for charging station

$$En_{c,i,t}^{GTS} = En_{c,i,t}^{GTBT} + En_{c,i,t}^{GTUC} + En_{c,i,t}^{GTFW} + \sum_{\forall c}^{at\,s} En_{c,i,t}^{GTC} - \left(\begin{array}{c} En_{c,i,t}^{BTTG} + En_{c,i,t}^{UCTG} + En_{c,i,t}^{FWTG} \\ + \sum_{\forall c}^{at\,s} En_{c,i,t}^{CTG} \end{array} \right) \quad (17.1)$$

Total energy supplied from the stations to the grid is shown in Eq. (17.2):

$$En_{c,i,t}^{STG} = En_{c,i,t}^{BTTG} + En_{c,i,t}^{UCTG} + En_{c,i,t}^{FWTG} + \sum_{\forall c}^{at\,s} En_{c,i,t}^{CTG} - \left(\begin{array}{c} En_{c,i,t}^{GTBT} + En_{c,i,t}^{GTUC} + En_{c,i,t}^{GTFW} \\ + \sum_{\forall c}^{at\,s} En_{c,i,t}^{GTS} \end{array} \right) \quad (17.2)$$

Price of energy supplied to the grid is shown in Eq. (17.3):

$$
\begin{aligned}
EP_{c,i,t}^{STG} = & \left(En_{c,i,t}^{BTTG} * EP_{BT,i,c,t}^{Dis} \right) + \left(En_{c,i,t}^{UCTG} * EP_{UC,i,c,t}^{Dis} \right) + \left(En_{c,i,t}^{FWTG} * EP_{FW,i,c,t}^{Dis} \right) \\
& + \left(\sum_{\forall c}^{at\,s} En_{c,i,t}^{CTG} * EP_{EV,i,c,t}^{Dis} \right) + \left(En_{c,i,t}^{NRTG} * EP_{NR,c,i,t}^{Chg} \right) + \left(En_{c,i,t}^{PVTG} * EP_{PV,c,i,t}^{Chg} \right) \\
& + \left(En_{c,i,t}^{FCTG} * EP_{FC,c,i,t}^{Chg} \right) + \left(En_{c,i,t}^{WTTG} * EP_{WT,c,i,t}^{Chg} \right) - \left(
\begin{array}{l}
\left(En_{c,i,t}^{GTBT} * EP_{BT,i,c,t}^{Chg} \right) \\
+ \left(En_{c,i,t}^{GTUC} * EP_{UC,i,c,t}^{Chg} \right) \\
+ \left(En_{c,i,t}^{GTFW} * EP_{FW,i,c,t}^{Chg} \right) \\
+ \left(\sum_{\forall c}^{at\,s} En_{c,i,t}^{GTC} * EP_{EV,i,c,t}^{Chg} \right)
\end{array}
\right) \quad (17.3)
\end{aligned}
$$

Price of energy supplied from the grid is shown in Eq. (17.4):

$$
\begin{aligned}
EP_{c,t}^{GTS} = & \left(En_{c,i,t}^{GTBT} * EP_{BT,i,c,t}^{Chg} \right) + \left(En_{c,i,t}^{GTUC} * EP_{UC,i,c,t}^{Chg} \right) + \left(En_{c,i,t}^{GTFW} * EP_{FW,i,c,t}^{Chg} \right) \\
& + \left(\sum_{\forall c}^{at\,s} En_{c,i,t}^{GTC} * EP_{EV,i,c,t}^{Chg} \right) - \left(
\begin{array}{l}
\left(En_{c,i,t}^{BTGGT} * EP_{BT,i,c,t}^{Dis} \right) + \left(En_{c,i,t}^{UCTG} * EP_{UC,i,c,t}^{Dis} \right) \\
+ \left(En_{c,i,t}^{FWTG} * EP_{FW,i,c,t}^{Dis} \right) + \left(\sum_{\forall c}^{at\,s} En_{c,i,t}^{CTG} * EP_{EV,i,c,t}^{Dis} \right) \\
+ \left(En_{c,i,t}^{NRTG} * EP_{NR,c,i,t}^{Chg} \right) + \left(En_{c,i,t}^{PVTG} * EP_{PV,c,i,t}^{Chg} \right) \\
+ \left(En_{c,i,t}^{FCTG} * EP_{FC,c,i,t}^{Chg} \right) + \left(En_{c,i,t}^{WTTG} * EP_{WT,c,i,t}^{Chg} \right)
\end{array}
\right) \quad (17.4)
\end{aligned}
$$

The trade agents will decide and negotiate the selling price, while controllers will decide the power transfer. The coordination between trade agents and controllers will optimize power flow based on negotiated and coordinated profits. Figure 17.6 shows the trade agents for the nearest stations CS1, CS2, and CS3, communicating with the EV trade agent. The communication is based on obtaining charging price for EV based on price calculation within each station, in view of available energy resources and storage. Minimum and maximum bidding values are defined for EV trade agents and the agents of all stations. Optimization techniques are applied to

Fig. 17.6 EV transactive charging

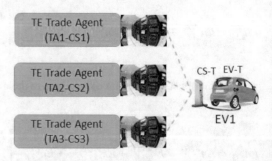

achieve an acceptable price, and EV will select the best bid. The trade agents for each component, battery, flywheel, ultracapacitor, PV, wind turbine, fuel cell, and nuclear reactor will coordinate to achieve local optimum performance with coordination and negotiation with other agents. The multi-investor model will allow fair negotiation for profitable energy and charging infrastructures. For example, investor-1 could hold ownership of PV, while investor-1 and investor-2 might own NR. The negotiation among trade agents will ensure fair and profitable transactive energy and charging station.

The estimation of energy price for each component will include lifecycle cost, including environmental impacts, and priority index of the energy system, as per national policies' user preference. For example, if the energy price supplied by PV is equivalent to the energy price supplied from the grid, then the priority index is used to decide energy resources. The priority index of PV is higher than the priority index of the grid. Hence, the trade will select PV to supply energy to the charging station. The trading agent will send instructions to the charging controller to enable charging from PV.

17.4 Transactive Mobility

There are continuous improvements to the mobility infrastructures where new technologies are adopted such as connected and autonomous vehicles (CAVs), drones, scooters, and flying cars. Then progress in transportation electrification is further enabling fast-charging and ultra-fast-charging infrastructures, with wireless capabilities. The business models associated with mobility are also changing to enable smooth movement of persons and goods. Ownership of vehicles might be complemented with flexible transit network and shared transportation. The investment models are also changing in view of these new mobility strategies. The lack of digital infrastructure to support the enhanced mobility networks is limiting the fast progress and transition. The proposed framework shows the developed digital infrastructure to support transactive mobility where the value of mobility is estimated and transactions are processed from mobility service providers to end-users. Figure 17.7 shows the digital infrastructure and associated mobility model that include end-users from the public, transit companies and municipalities, drivers, service providers, utilities for energy supply, and investors. The proposed digital infrastructure will include scenario modeling, co-simulation, real-time data analytics, AI optimization algorithms, and financial and cost monitoring. The proposed platform will support policy development, city planning, transit planning, and goods shipping and movement including maritime, road, rail, and air shipping.

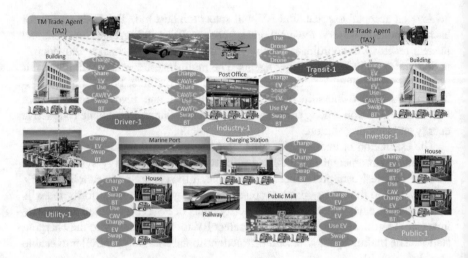

Fig. 17.7 Transactive mobility architecture

17.5 Summary

This chapter presented a framework for transactive energy for transactive charging. Energy systems are defined for the charging station, including renewable energy resources and energy storage. The hybrid energy storage system is integrated to ensure diversity and high energy storage performance. Trade agents are defined for each component, such as EV, PV, wind turbine, fuel cell, energy storage, nuclear reactor, and the charging unit. The cost model is defined for each component to ensure an accurate estimate of energy price for each component. Some components offer bidirectional energy flow, such as a battery. Other components offer unidirectional energy flow, such as PV. Investment is encouraged to support the establishment of charging infrastructures with more penetration and deployment.

Acknowledgments Authors would like to thank the members in the Smart Energy Systems Lab (SESL), in particular Dr. Yasser Elsayed for his support to the work.

References

1. H.R. Bokkisam, M.P. Selvan, Effective community energy management through transactive energy marketplace, Comput. Electr. Eng., **93**, 107312 (2021)
2. W. Liu, J. Zhan, C.Y. Chung, A novel transactive energy control mechanism for collaborative networked microgrids, IEEE Trans. Power Syst. **34**(3), 2048–2060 (2019)
3. Q. Huang, T.E. McDermott, Y. Tang, A. Makhmalbaf, D.J. Hammerstrom, A.R. Fisher, L.D. Marinovici, T. Hardy, Simulation-based valuation of transactive energy systems. IEEE Trans. Power Syst. **34**(5), 4138–4147 (2019)

4. M. Daneshvar, B. Mohammadi-Ivatloo, M. Abapour, S. Asadi, R. Khanjani, Distributionally robust chance-constrained transactive energy framework for coupled electrical and gas microgrids. IEEE Trans. Ind. Electron. (1982) **68**(1), 347–357 (2021)

5. Z. Liu, Q. Wu, M. Shahidehpour, C. Li, S. Huang, W. Wei, Transactive real-time electric vehicle charging management for commercial buildings with PV on-site generation. IEEE Trans. Smart Grid **10**(5), 4939–4950 (2019)

6. A. Mohammad, R. Zamora, T.T. Lie, Transactive energy management of PV-based EV integrated parking lots. IEEE Syst. J. **15**(4), 5674–5682 (2021)

7. Z. Liu, Q. Wu, K. Ma, M. Shahidehpour, Y. Xue, S. Huang, Two-stage optimal scheduling of electric vehicle charging based on transactive control. IEEE Trans. Smart Grid **10**(3), 2948–2958 (2019)

8. Z. Wu, B. Chen, Distributed electric vehicle charging scheduling with transactive energy management. Energies (Basel) **15**(1), 163 (2021)

9. D. Kudathanthirige, R. Shrestha, A. Baduge, G. Amarasuriya, Max-min fairness optimal rate-energy trade-off of SWIPT for massive MIMO downlink. IEEE Commun. Lett. **23**(4), 688–691 (2019)

10. H. Zhang, Y. Wang, X. Zhu, Y. Guo, The impact of energy trade patterns on CO2 emissions: An emergy and network analysis. Energy Econ. **92**, 104948 (2020)

11. H.B. Sakal, Turkey's energy trade relations with europe: The role of institutions and energy market, energy & environment (Essex, England) **32**(7), 1243–1274 (2021)

12. C. Duan, B. Chen, Energy–water nexus of international energy trade of China. Appl. Energy **194**, 725–734 (2017)

Correction to: Fast Charging and Resilient Transportation Infrastructures in Smart Cities

Hossam A. Gabbar

Correction to:
H. A. Gabbar, *Fast Charging and Resilient Transportation Infrastructures in Smart Cities*, https://doi.org/10.1007/978-3-031-09500-9

In Chapters 3, 4, 5, 6, 7, and 16, the chapter author names were mentioned inadvertently in the footnotes. These have now been deleted and the author names have been linked to the author group citation in the aforementioned chapters.

The updated original versions of the chapters can be found at
https://doi.org/10.1007/978-3-031-09500-9_3
https://doi.org/10.1007/978-3-031-09500-9_4
https://doi.org/10.1007/978-3-031-09500-9_5
https://doi.org/10.1007/978-3-031-09500-9_6
https://doi.org/10.1007/978-3-031-09500-9_7
https://doi.org/10.1007/978-3-031-09500-9_16

Index

© Springer Nature Switzerland AG 2022
H. A. Gabbar, *Fast Charging and Resilient Transportation Infrastructures in Smart Cities*, https://doi.org/10.1007/978-3-031-09500-9

Printed in the United States
by Baker & Taylor Publisher Services